Design

A Practical Guide to the RIBA Plan of Work 2013
Stages 2 and 3

Tim Bailey

RIBA **Publishing**

Contents

Foreword
v

Series editor's foreword
vi–vii

The author
viii

Acknowledgements
viii

The series editor
ix

The scenarios
x

The in-text boxed features
xi

The RIBA Plan of Work 2013
xii–xiii

Introduction
xiv–xviii

01 **STARTING STAGES 2 AND 3**
1–25

02 **STAGE 2 CONCEPT DESIGN**
27–77

03 **STAGE 3 DEVELOPED DESIGN**
79–127

04 **CONCLUSION**
129–139

Plan of Work glossary
140–143

Index
144–147

© RIBA Enterprises, 2015

Published by RIBA Publishing,
66 Portland Place, London, W1B 1AD

ISBN 978 1 85946 571 4

Stock code 83010

The right of Tim Bailey to be identified as the Author of this Work has been asserted in accordance with the Copyright, Design and Patents Act 1988, sections 77 and 88.

All rights reserved. No part of this publication may be reproduced, stored in a retrieval system, or transmitted, in any form or by any means, electronic, mechanical, photocopying, recording or otherwise, without prior permission of the copyright owner.

British Library Cataloguing in Publications Data
A catalogue record for this book is available from the British Library.

Commissioning editor: Sarah Busby
Production: Michèle Woodger
Designed and typeset by: Alex Lazarou
Printed and bound by: CPI
Cover image: © iStock/Franck-Boston

While every effort has been made to check the accuracy and quality of the information given in this publication, neither the Author nor the Publisher accept any responsibility for the subsequent use of this information, for any errors or omissions that it may contain, or for any misunderstandings arising from it.

RIBA Publishing is part of RIBA Enterprises Ltd.
www.ribaenterprises.com

Foreword

Charles Eames once reflected that 'without problems there would be no design'. So the job of design is to solve problems.

The RIBA Plan of Work has always looked to expand upon Eames perceptive and pithy declaration whilst remaining true to its principle. It offers a framework for the process and presentation of design, from inception through to delivery, whilst avoiding constraint. Solutions can be tailored to suit and decisions brought forward (I anticipate exactly that for planning submissions) or deferred as appropriate. For all the above reasons it has provided an important structure for the iterations of design that is understood by clients, consultants and architects alike.

In this context the need for the new RIBA Plan of Work can be explained by a structural change in the nature of how we perceive problems and how we create and record design as the response. The greatest single structural change to how we design, unsurprisingly in our audited world, is the rising importance of assessing risk. Risk and its identification, recording, transfer and ideally elimination through the design process is the new key design problem. This is reflected in construction by the transfer of risk from client to contractor and the consequent rise of design and build contracts.

Necessary, and as well received as it may be, the new Plan of Work, by changing fields of reference familiar to clients and consultants alike, creates at least initially, the problem of uncertainty. And it is this uncertainty that this new publication, one of a series, sets out to address in extended detail. To that end I believe it will be a useful, graphic and, through its illustrated case studies, a well-grounded and detailed exploration and explanation of the complex process of recording problems and their integrated design solution: from idea to detail without forgetting delight.

As another great designer, Raymond Loewy once remarked of design 'you can never leave well enough alone' and with this in mind this well-considered book will prove useful to designers in both answering the concerns and justifying the benefits of the new design for the RIBA Plan of Work.

Simon Allford
Allford Hall Monaghan Morris

Series editor's foreword

The RIBA Plan of Work Stage Guides are a crucial accompaniment to the RIBA Plan of Work 2013. The plan's format cannot communicate or convey the detail behind every term in the plan and this series provides essential guidance by considering, in depth, the reasoning and detail behind many new and reinvigorated subjects linking these to practical examples. The series is comprised of three titles which each concentrate on distinct stages in the Plan of Work. The first is Briefing by Paul Fletcher and Hilary Satchwell which covers Stages 7, 0 and 1. The second is Design by Tim Bailey and this covers Stages 2 and 3. The third is Construction by Phil Holden and covers Stages 4, 5 and 6. Subjects explored include how to assemble the most appropriate and effective project team and how to develop the best possible brief. The series also considers how to deal with the cultural shifts arising from a shift from "analogue" to transformational "digital" design processes as our industry begins to absorb the disruptive technologies that are changing many different and diverse sectors beyond recognition.

The RIBA Plan of Work 2013 drives a shift towards richer and bigger data which can be harnessed to create better whole life outcomes and thus significant additional benefits to clients and users. The first book in the series, Briefing, considers how the new project stages (0 and 7) will add value over the lifetime of a project as greater emphasis is placed on more resilient designs where whole life considerations are embedded into the early design stages. With this in mind the series emphatically starts with Stage 7 placing emphasis on the importance of learning from previous projects via feedback and in the future via data analytics. This initial chapter also sets out how post occupancy and building performance evaluations can be harnessed to inform the Business Case during Stage 0 underlining that big data will provide paradigm shifts in how to extract feedback from newly completed or existing projects, including historic buildings, to help better decision making in the early project stages.

More specifically, Paul and Hilary's book considers new Stage 7 to 0 activities that will result in exciting new services in the future. These will ensure that the client's brief is robust and properly considered providing the best possible platform for the design stages. This publication also considers the importance of site appraisals at Stage 0 and how Feasibility Studies can assist and add value at Stage 1 to the briefing process before the design process commences in earnest at Stage 2. In every stage there is added emphasis around Information Exchanges and the importance of considering who does what when at the outset of a project.

Although the core design stages (2 and 3) have not significantly changed, Tim Bailey's book, Design, looks at how they might be adjusted and better focused to provide greater client emphasis at Stage 2 allowing the lead designer to take centre stage at Stage 3. During this stage greater emphasis is placed on the production of a co-ordinated design: the design team should be focused on the work required to verify that the Concept Design is robust and suitable for making a Planning application. In both stages new methods of communicating the progressing design create exciting new opportunities but at the same time require an examination of how to effectively manage the design process using tools such as the Design Programme to manage what is an iterative process.

Finally, Phil Holden's book, Construction, considers the complexities of Stage 4 which is "sliced and diced" in different ways depending on the procurement route and the extent of design work undertaken by the specialist subcontractors employed by the contractor. He considers how the Design Programme for this change might alter to reflect different procurement routes and how this stage typically overlaps with construction (Stage 5). Handing over projects is becoming increasing complex and users now realise that the handover process can impact on successful operation and use of their buildings. Phil considers how the handover process is changing, placing greater emphasis on the user's needs. His Stage 6 narrative considers how building contracts might adapt to this new environment placing greater emphasis on whole life matters including achieving better project outcomes rather than focusing on the closure solely of contractual matters and construction defects.

Five project scenarios weave through the series providing some practical examples of how the different stages of the plan of work might be interpreted on projects of differing scales, sectors, complexity using different procurement routes, providing a consistent thread through all of the books.

In summary, the series provides excellent additional guidance on how to use the RIBA Plan of Work 2013 allowing anyone involved in the built environment to understand and use the plan more effectively with the goal of achieving better whole life outcomes.

The author

Tim Bailey is founder of xsite architecture, established in 2000 and based in Newcastle upon Tyne. Tim has been an architect since 1992, and has been involved with a diverse range of realised projects throughout north-east England and London. Born in the north-east and graduating from Newcastle University, he made a positive decision to stay in the region and create a practice that had a strong design ethos underpinning a realistic commercial approach to projects. In addition, he has developed a strong relationship with the arts sector, working collaboratively with and for arts clients as well as in the retail, leisure, residential and commercial sectors.

He is currently a Regionally Elected member of the RIBA National Council, RIBA Board member without portfolio and member of the RIBA Practice and Profession Committee.

Acknowledgements

There are always people to thank for the twists and turns that got us to the place we are in. Thanks to them all but in particular thanks to Dale Sinclair for asking me to take on this book, it has been very enjoyable. Thanks to my fellow authors in this part of the Plan of Work series, Hilary Satchwell, Paul Fletcher and Phil Holden, for probing and challenging the format of the series and in consequence this book into shape. Thanks to Sarah Busby for the guidance, gentle cajoling and support throughout.

A big love and thanks to Ruth, Zac and Thea for their holiday patience, lost evenings and timely groundings. Payback starts now!

The series editor

Dale Sinclair is Director of Technical Practice for AECOM's architecture team in EMEA. He is an architect and was previously a Director of Dyer and an Associate Director at BDP. He has taught at Aberdeen University and the Mackintosh School of Architecture and regularly lectures on BIM, design management and the RIBA Plan of Work 2013. He is passionate about developing new design processes that can harness digital technologies, manage the iterative design process and improve design outcomes.

He is currently the RIBA Vice President, Practice and Profession, a trustee of the RIBA Board, a UK board member of BuildingSMART and a member of various Construction Industry Council working groups. He was the editor of the BIM Overlay to the Outline Plan of Work 2007, edited the RIBA Plan of Work 2013 and was author of its supporting tools and guidance publications: Guide to Using the RIBA Plan of Work 2013 and Assembling a Collaborative Project Team.

The scenarios

Throughout the series five projects of different scale, sector and complexity have been used to illustrate the practical impact of the RIBA Plan of Work 2013. These look at how different projects may need to deal differently with a range of issues that could arise. These are not intended to be definitive examples of what to do, or what not to do, but to aid understanding of the plan of work and how different approaches may be adopted at each stage to support better project outcomes. They are:

- **Scenario A: An extension to a four-bedroom house in a rural location.** This project is for a private client and has a budget of £250k. The design team have been selected by recommendation from friends and are appointed to help the client develop the brief. The chosen procurement route is the traditional procurement of a contractor by the client.
- **Scenario B: A small scale housing development for a local developer on the outskirts of a large city.** The value of the project is £1.5million and the client is a small but well established family business. Both the design team and the contractor are to be selected by informal tender with previous experience and pricing core evaluation factors. The procurement route is also traditional.
- **Scenario C: The refurbishment of a teaching building for a University which has a large portfolio of buildings.** The value of the project is between £5million – £6million. The design team are selected following a mini competition. The procurement route is single stage design and build with the design team being novated to the contractor.
- **Scenario D: A new central library for a medium sized Local Authority.** Following the development of the brief including Feasibility Studies produced by a directly appointed team on the Council's Consultant Framework the project is tendered to select the design team for the next stages. The contractor is to be selected following a two stage design and build process and will appoint their own design team. The original design team is to be retained by the council as advisors.
- **Scenario E: A large office scheme for a high tech internet based company wanting to establish themselves as major players in the industry with a high profile new base.** Valued at £18 million - £20 million this project is procured using a management form of contract due to the urgent need to occupy the building.

At the end of each Stage in the book there is a status check on the five projects where the impact of the work and decisions made during that stage are illustrated. Within each chapter these scenarios are used to identify key points and examples.

The in-text boxed features

We have also included several in-text boxed features to enhance your understanding of the Plan of Work stages and their practical application.

The following key will explain what each icon means and why each feature is useful to you:

The 'Example' feature explores an example from practice, either real or theoretical, and often utilizing the project scenarios.

The 'Hints and Tips' feature dispenses pragmatic advice and highlights common problems and solutions.

The 'Definition' feature explains key terms in more detail.

DESIGN
A PRACTICAL GUIDE TO RIBA PLAN OF WORK 2013
STAGES 2 AND 3

RIBA Plan of Work 2013

The **RIBA Plan of Work 2013** organises the process of briefing, designing, constructing, maintaining, operating and using building projects into a number of key stages. The content of stages may vary or overlap to suit specific project requirements.

Tasks ▼	**0** Strategic Definition	**1** Preparation and Brief	**2** Concept Design	**3** Developed Design
Core Objectives	Identify client's **Business Case** and **Strategic Brief** and other core project requirements.	Develop **Project Objectives**, including **Quality Objectives** and **Project Outcomes**, **Sustainability Aspirations**, **Project Budget**, other parameters or constraints and develop **Initial Project Brief**. Undertake **Feasibility Studies** and review of **Site Information**.	Prepare **Concept Design**, including outline proposals for structural design, building services systems, outline specifications and preliminary **Cost Information** along with relevant **Project Strategies** in accordance with **Design Programme**. Agree alterations to brief and issue **Final Project Brief**.	Prepare **Developed Design**, including coordinated and updated proposals for structural design, building services systems, outline specifications, **Cost Information** and **Project Strategies** in accordance with **Design Programme**.
Procurement *Variable task bar	Initial considerations for assembling the project team.	Prepare **Project Roles Table** and **Contractual Tree** and continue assembling the project team.	←- The procurement strategy does not fundamentally alter the progression of the design or the level of detail prepared at	a given stage. However, **Information Exchanges** will vary depending on the selected procurement route and **Building Contract**. A bespoke RIBA Plan of Work
Programme *Variable task bar	Establish **Project Programme**.	Review **Project Programme**.	Review **Project Programme**.	←- The procurement route may dictate the **Project Programme** and result in certain stages overlapping
(Town) Planning *Variable task bar	Pre-application discussions.	Pre-application discussions.	←- Planning applications are typically made using the Stage 3 output.	A bespoke **RIBA Plan of Work 2013** will identify when the
Suggested Key Support Tasks	Review **Feedback** from previous projects.	Prepare **Handover Strategy** and **Risk Assessments**. Agree **Schedule of Services**, **Design Responsibility Matrix** and **Information Exchanges** and prepare **Project Execution Plan** including **Technology** and **Communication Strategies** and consideration of **Common Standards** to be used.	Prepare **Sustainability Strategy, Maintenance and Operational Strategy** and review **Handover Strategy** and **Risk Assessments**. Undertake third party consultations as required and any **Research and Development** aspects. Review and update **Project Execution Plan**. Consider **Construction Strategy**, including offsite fabrication, and develop **Health and Safety Strategy**.	Review and update **Sustainability, Maintenance and Operational** and **Handover Strategies** and **Risk Assessments**. Undertake third party consultations as required and conclude **Research and Development** aspects. Review and update **Project Execution Plan**, including **Change Control Procedures**. Review and update **Construction** and **Health and Safety Strategies**.
Sustainability Checkpoints	**Sustainability Checkpoint — 0**	**Sustainability Checkpoint — 1**	**Sustainability Checkpoint — 2**	**Sustainability Checkpoint — 3**
Information Exchanges (at stage completion)	**Strategic Brief**.	**Initial Project Brief**.	**Concept Design** including outline structural and building services design, associated **Project Strategies**, preliminary **Cost Information** and **Final Project Brief**.	**Developed Design**, including the coordinated architectural, structural and building services design and updated **Cost Information**.
UK Government Information Exchanges	Not required.	Required.	Required.	Required.

*Variable task bar – in creating a bespoke project or practice specific RIBA Plan of Work 2013 via www.ribaplanofwork.com a specific bar is selected from a number of options.

THE RIBA PLAN OF WORK 2013

The **RIBA Plan of Work 2013** should be used solely as guidance for the preparation of detailed professional services contracts and building contracts.

www.ribaplanofwork.com

4 Technical Design	5 Construction	6 Handover and Close Out	7 In Use
Prepare **Technical Design** in accordance with **Design Responsibility Matrix** and **Project Strategies** to include all architectural, structural and building services information, specialist subcontractor design and specifications, in accordance with **Design Programme**.	Offsite manufacturing and onsite **Construction** in accordance with **Construction Programme** and resolution of **Design Queries** from site as they arise.	Handover of building and conclusion of **Building Contract**.	Undertake **In Use** services in accordance with **Schedule of Services**.
2013 will set out the specific tendering and procurement activities that will occur at each stage in relation to the chosen procurement route. →	Administration of **Building Contract**, including regular site inspections and review of progress.	Conclude administration of **Building Contract**.	
or being undertaken concurrently. A bespoke **RIBA Plan of Work 2013** will clarify the stage overlaps.	The **Project Programme** will set out the specific stage dates and detailed programme durations. →		
planning application is to be made. →			
Review and update **Sustainability, Maintenance and Operational** and **Handover Strategies** and **Risk Assessments**. Prepare and submit Building Regulations submission and any other third party submissions requiring consent. Review and update **Project Execution Plan**. Review **Construction Strategy**, including sequencing, and update **Health and Safety Strategy**.	Review and update **Sustainability Strategy** and implement **Handover Strategy**, including agreement of information required for commissioning, training, handover, asset management, future monitoring and maintenance and ongoing compilation of **'As-constructed' Information**. Update **Construction** and **Health and Safety Strategies**.	Carry out activities listed in **Handover Strategy** including **Feedback** for use during the future life of the building or on future projects. Updating of **Project Information** as required.	Conclude activities listed in **Handover Strategy** including **Post-occupancy Evaluation**, review of **Project Performance**, **Project Outcomes** and **Research and Development** aspects. Updating of **Project Information**, as required, in response to ongoing client **Feedback** until the end of the building's life.
Sustainability Checkpoint — 4	Sustainability Checkpoint — 5	Sustainability Checkpoint — 6	Sustainability Checkpoint — 7
Completed **Technical Design** of the project.	**'As-constructed' Information**.	Updated **'As-constructed' Information**.	**'As-constructed' Information** updated in response to ongoing client **Feedback** and maintenance or operational developments.
Not required.	Not required.	Required.	As required.

© RIBA

Introduction

The RIBA Plan of Work 2013

In the overview framework of the RIBA Plan of Work 2013, the eight sequential stages, Stages 0–7, that represent the briefing, designing, constructing, maintaining, operating and in use stages of a project are arranged horizontally so that the eight task bars, arranged vertically, that explain fixed and variable activity – for example, Procurement, Programme and (Town) Planning – can describe relevant and key activities under each of the stages. This framework establishes, for the project team, a robust management tool of the design processes in any project, and, with the variable task bars, has the flexibility to adjust to practice-specific or project-specific circumstances very easily.

Introducing the Stage Guides series

This guide is the middle book of three that look in detail at the RIBA Plan of Work 2013 stages. It discusses Stage 2 Concept Design and Stage 3 Developed Design.

The first book in the series covers the beginnings of a project, and has chosen to start with Stage 7 In Use. It does so in order to establish the case that when a client might consider a project, they often do so on the basis of the accommodation they are in or their experiences of previous projects that they have been involved with. It may be the fact that their building is too small, too large or too expensive to run that prompts the line of enquiry that results in a project. At this very early stage, there is no presumed solution (for example, a refurbishment, retrofit or new build), and the initial stages are set up to establish all the relevant criteria that will inform future design stages. Following that starting point to the first guide book is Stage 0 Strategic Definition, when a client considers the existing data set from the Stage 7 information available. They assemble the first members of a project team, attempt to define the project within a Strategic Brief, establish a Project Programme and, for some projects, may also establish a Project Budget. Immediately prior to the activity discussed in this book comes Stage 1 Preparation and Brief, wherein Feasibility Studies are undertaken to test the brief. During Stage 1 as the project brief starts to emerge, it is possible to identify the key Project Strategies that will be developed during Stages 2 and 3. Project documentation begins to be produced prior to collation into an Information Exchange at the end of Stage 1.

The complexity of the project being undertaken will determine the levels of detail produced at this very early stage in its life, but it should include the organisational structure for the project. For example, a thorough understanding of the client's Business, or the circumstances that have prompted a capital project, will help determine a project-management structure, existing or proposed site appraisals to establish the appropriate scope of work, and the appointment of the design team necessary for the project. This information will be recorded in the initial Project Execution Plan (PEP), which will also include a Design Responsibility Matrix that establishes who is responsible for the production of each element of design work and a Communication Strategy that allows a total understanding by the whole project team of how project data is to be transferred, to whom and in what timeframes. Other relevant strategies will be explained during the course of that book.

The third book in the series covers RIBA Plan of Work 2013 Stage 4 Technical Design, Stage 5 Construction and Stage 6 Handover and Close Out. The stage titles in this book are fairly self-explanatory, covering the period of time in any building project that deals with the practical and physical arrival of the design and the beginning of its use by the intended occupants.

Stage 7 captures the building project in use; it is when feedback and evaluation data on the post-occupancy performance of the building is gathered, building performance evaluated, maintenance regimes tested and modified, and the 'As-constructed' Information is kept up to date with any change during the building's use. Hopefully, it is clear that it makes sense, within this Stage Guide series, to discuss Stage 7 information at the beginning of the process, despite the fact that this stage is placed at the end of the RIBA Plan of Work 2013 framework, as it will have a bearing on the assembly of criteria for the building project about to be embarked upon.

This guide, together with the other titles in the series, aims to populate for the reader the RIBA Plan of Work 2013 framework by exploring the activities behind the identified tasks at each stage. By providing a sense of where each stage starts and ends, how to manage and complete the stages, and by highlighting the importance of the collaborative, iterative and methodical development of design ideas and solutions, it is intended that design team members will discover how to work inside the RIBA Plan of Work 2013 framework and the resulting benefits in realising a coordinated and tangible building project.

DESIGN
A PRACTICAL GUIDE TO RIBA PLAN OF WORK 2013
STAGES 2 AND 3

Stage	7 In Use	0 Strategic Definition	1 Preparation and Brief	2 Concept Design	3 Developed Design	4 Technical Design	5 Construction	6 Handover & Close Out	7 In Use
Review/ analysis	**Book 1: Briefing** — In Use Data, Strategic Brief, Concept Design								
Design/ synthesis				**Book 2: Design** — Final Project Brief, Developed Design					
Delivery/ process						**Book 3: Construction** — Technical Design, Construction, Handover			

0.1 Each book in the Stage Guides series mapped against RIBA Plan of Work 2013.

What is this book about?

It is no accident that Stages 2, 3 and 4 of the RIBA Plan of Work 2013 have the word 'Design' in their titles. These are the stages at which the design team assembled by the client for their building project play their most important role. It is during Stage 2 Concept Design and Stage 3 Developed Design that a building project takes form, from an idea sparked by the Initial Project Brief. Site and organisational data and information collected during the briefing stages are turned into a visualised, coordinated building-project proposal that is ready to be tested against external opinion. This may be in the form of the planning authority, the neighbourhood or general public – or even, perhaps, the press – as well as any pertinent legislation that controls development in the project location. An equally potent opinion will be formed by 'the market', whether that be commercial viability or market demand for the project type and funding regime. The application of skill, experience and imagination from the design team during these early stages of a project's life can give rise to powerful and lasting statements of society's civic, cultural and commercial ambitions in any period of history, and the rigour with which they are delivered plays a key part in their quality. The RIBA Plan of Work 2013 is intended to provide the organisational framework and surrounding guidance to illustrate how that rigour can be applied to all projects irrespective of size, complexity or cost.

How are project resources decided for design stages?

Arguably, the most important resourcing decisions are taken within the design team in the early part of Stage 2. Who will or should work on the project within each discipline? Should they be available through its entire programme, in order to provide consistency of concept, contact and communication? Is there a client sensitivity behind the brief that requires a particular skill set or resource type, and will that change through the course of the whole project? What is the level of design talent or technical competency required to derive the best result from the process?

RIBA Plan of Work 2013 Stage 1 will have left plenty of clues to the answers to these questions, together with evidence of the early conversations with clients and their agents about what their ambitions and expectations for the project are. Depending on the size of the project and the relative size of the design companies involved, the decisions about how to resource the early stages of one project will be very different from those of another, but the rigour behind the assembly of the Initial Project Brief should be the same in every case. The appointment process might be by fee bid, design competition, competitive interview or personal recommendation. Each of these situations will result in some members of the design team having heard more about the project than others. Key to getting the project started successfully as design team appointments are confirmed, is affording the rest of the design team the opportunity to hear those client ideas and ambitions, and the relevant project briefing information, for themselves. A great deal of the confidence and inspiration that can 'carry' the first period of any commission will come from this first-hand experience.

Stages 2 and 3 under the microscope

This book is focused on RIBA Plan of Work 2013 Stages 2 and 3. These stages cover the period of the project when a building design emerges from a written brief into a fully declared building proposal. This is a very creative part of the process of realising buildings, which benefits hugely from the rigour imposed by the Plan of Work. Below is a short insight into what the chapters of this book cover:

- At the outset of the book, Chapter 1 looks at the two stages and explains what Concept Design and Developed Design are. How do design teams plan the resources needed to tackle these stages, and who should lead the design team? What preparation do design teams need to start Stage 2?
- In Chapter 2, the Concept Design stage is unpacked. How do design concepts 'arrive', and how does the essentially creative process of designing architecture respond to being defined by a framework like the Plan of Work 2013? Who does what during this stage, and how are the work outputs during Stage 2 represented and assembled at stage completion?
- Chapter 3 covers the Developed Design stage. This stage is characterised by the declaration of the design and the production of a great deal of detailed information in the form of reports, drawings and models. The chapter looks at what form the information will be in, what actions are taken with it and how its decisions impact on Project Strategies and other Project Information.
- Chapter 4 summarises the guide, and looks briefly at what to expect in the next RIBA Plan of Work 2013 stage: Stage 4 Technical Design.

At the end of the book, there is a glossary of key terms referenced in the book and relating back to those included in the Plan of Work 2013.

CHAPTER 01

STARTING STAGES 2 AND 3

CHAPTER 01

OVERVIEW

This second guide in the series unlocks the RIBA Plan of Work 2013 framework for the principal design stages: Stages 2 (Concept Design) and 3 (Developed Design). This chapter briefly discusses what these stages cover in the context of all eight Plan of Work stages. The chapter identifies what comes before Stage 2, and what lies ahead after Stage 3. A short section follows on the importance of the review and development of Project Strategies and the key role of task bars in organising this information within the Plan of Work framework. The subject of design leadership and how to resource these stages is also mentioned as being a major factor in the successful completion of Plan of Work projects.

WHAT IS A CONCEPT DESIGN?

The expression 'Concept Design' could do with some explanation in the context of the RIBA Plan of Work 2013 Stage 2. It is important that all members of the project team understand what is meant by Concept Design, as there is often a misconception that it relates to the first idea only. But more critically than explaining how the design team needs to develop the building project through RIBA Plan of Work 2013 Stage 2, is the need for the client to know what to expect from the design team at the point where they declare a particular Concept Design as the right solution for the Initial Project Brief.

A Concept Design is the expression of a central and core idea, around which the constituent parts of a project can be based. For some projects, this becomes the one-line description that inspires or captures the essence and nature of the proposal. Often, and sometimes famously, the concept is referred to as having been captured on the 'back of an envelope' or a napkin over dinner.

A SNAPSHOT – STAGE 2 CONCEPT DESIGN

The design team is to create a design response to the Initial Project Brief and Information Exchange from Plan of Work Stage 1. The Initial Project Brief should contain the Project Objectives, Project Outcomes, Project Budget, Project Programme and any site constraints that are understood to form part of Site Information. By the end of the Plan of Work Stage 2, the design team will have worked collaboratively to create a Concept Design that is set in an appropriate socioeconomic context and that has developed information, including Project Strategies, for the proposed scheme that gives it the prospect of developing into a realisable project.

In the context of the RIBA Plan of Work 2013, a Concept Design might start with an occurrence of that kind, but the whole of Stage 2 will explore the design much further – testing the context, site conditions and proposed accommodation of the building against the Initial Project Brief. The goal within this stage is to reach a point where what is presented as a Concept Design has a credible spatial, experiential and plan arrangement. The design team at this point will have the confidence to develop the outline structural design and building services systems and Sustainability Strategy, and, as will be demonstrated later, to produce a stage completion with a comprehensive set of Project Information to back up the sketch.

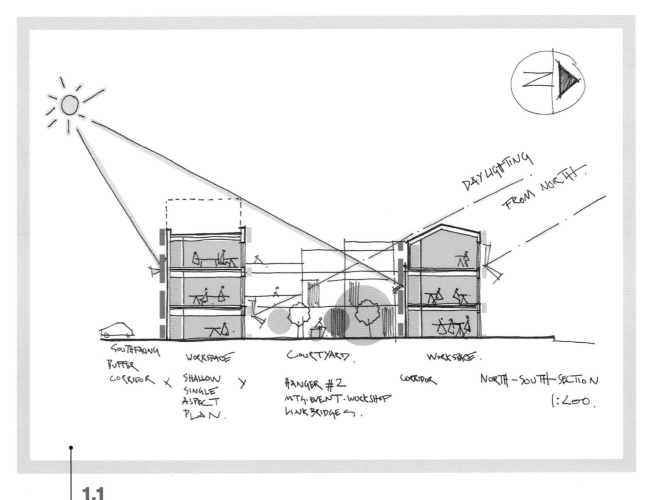

1.1
An example of a Concept Design sketch and drawing.

DESIGN
A PRACTICAL GUIDE TO RIBA PLAN OF WORK 2013
STAGES 2 AND 3

At the outset of Stage 2, the language used to discuss the emerging design might be vague and general. The purpose of developing the design using the RIBA Plan of Work 2013 is to create rigorous decision-making processes and investige each activity, to create a robust method and a solid basis for results. At the end of Stage 2, the design team will be able to present a design that has determined a real physicality, that describes the spatial arrangement of the proposal and a sense of context for the finished project. At this stage, it should be possible for the design team to have answered some fundamental questions about the project – for example:

- Does the brief establish a schedule of accommodation that satisfies its requirements?
- How much space should be allowed for the structural system?
- What materials will the building be made of?
- What insulation zone should be left to meet the established U-value criteria?
- What size should the plant room and major service-distribution routes be that are required for the building?

This level of investigation enables the creation of an Information Exchange at the conclusion of Stage 2 that uses an appropriate array of reports, drawings and visualisation tools, including 3D computer or physical modelling to illustrate the Concept Design.

WHAT IS A DEVELOPED DESIGN?

Similarly, it is useful to consider what is meant by a 'Developed Design', the title of RIBA Plan of Work 2013 Stage 3. A Developed Design in essence sets out a series of further tests for the assumptions built into the design during Stage 2. For the design team, this process is an opportunity to validate the Concept Design by undertaking coordination exercises between design team members, and by harnessing high-quality product and technical information that can assist in both design decision-making and the better understanding of the finished article. This whole process is ultimately carried out to achieve better certainty for the client on cost and programme criteria, but also to satisfy the client on the qualitative component of the project, which will be tested through the public consultation and planning processes that most often happen during this stage.

A SNAPSHOT – STAGE 3 DEVELOPED DESIGN

The design team refine their ideas by rigorously testing the design assumptions made at Stage 2. The design process at Stage 3 encompasses information gained from additional investigative work and involves making decisions about materials and construction products that support the Developed Design. The key task during Stage 3 is to assemble the developed designs from all design team members in order to provide the client with a coordinated design and robust cost plan. This level of confidence in the design allows the client to sign off the scheme and submit it for planning permission.

Some fundamental questions about the project arise during Stage 3 that need to be considered very carefully, as they often challenge decisions that become fixed. For example:

- Is the external materials palette chosen at Stage 2 capable of meeting all the technical requirements necessary, and does it have an effect on the cost plan?
- How do these material choices impact on the suite of Project Strategies that set out criteria for 'green' credentials, sustainability, embodied energy or maintenance cycles?

Similarly, a review of the Information Exchange from Stage 2 will need to check and update all the assumptions made at Concept Design stage that may be impacted by decisions emerging within Stage 3. For example:

- After review, are there spatial restrictions that impact on the choice of structural system?
- On review of the Sustainability Strategy, is the insulation zone agreed at Stage 2 insufficient to meet target U-value criteria with the least environmentally damaging insulation type?
- Have decisions taken about increasing the quantity of glazing in one part of the building affected the scale of the mechanical ventilation equipment installation, which in turn affects the size of the plant room required for the building?

These questions and reviews will prompt a series of discussions across the design team that result in confirming or adjusting allowances made at Stage 2. If there is significant impact on the Concept Design, the client will need to be consulted to agree the change of scope. Towards the end of Stage 3, these iterative design processes develop into a coordinated design agreed across the design team, which should not require further adjustment during RIBA Plan of Work 2013 Stage 4 Technical Design.

The principal aim of the Stage 3 process is to provide, upon its completion, an Information Exchange that delivers for both the project team and the project stakeholders a complete picture of how the building project will look, work and be used, and how the cost and programme components of the project have been developed and controlled to allow successful progression to RIBA Plan of Work 2013 Stages 4, 5 and 6.

STARTING STAGES 2 AND 3

1.2
An extract example of a Developed Design drawing.

PROJECT STRATEGIES AND DESIGN PROCESSES

A crucial part of both of these Plan of Work stages will be the involvement of a series of processes and devices for turning subjective views and choices into objective responses to the brief. While there is always room for a debate on a matter of style or taste, these must turn themselves into decisions that are made for rational and recordable reasons, with documentation that can explain – either as a narrative or as a prescription to the project brief – how the 'story' of the building unfolds.

Developing, and keeping under review at each stage, relevant Project Strategies is the main technique for creating a rationale for decision-making. Not only does the development of these strategies log the variety of options appraised and decisions taken, but, when supported by inclusive design team discussion, they also illustrate a properly carried-out component of the design process. It is important to recognise that each member of the design team has a critical role to play in developing Project Strategies in these early stages. The lead designer, for instance, will be able to coordinate diverse subjects and present these project complexities in an understandable way to the client. Strategy decisions made utilising the expert knowledge of the design team at these early stages will underpin the success of the later stages of the project and the building's performance in use.

SOME EXAMPLES OF PROJECT STRATEGIES

Sustainability Strategy – to include energy efficiency and renewable energy sources, building design life, whole-life costing, embodied-energy criteria, and reuse and recycle strategies.

Maintenance and Operational Strategy – to include maintenance and annual inspection cycles, including window-cleaning, requirements for keeping service records, plant and equipment replacement strategy, building-in-use recording, performance evaluation and security operation.

Health and Safety Strategy – to include project-specific health and safety information, CDM (Construction (Design and Management)) Regulations, accident-mitigation targets and project safety targets.

Risk Assessments – how to identify project risks and operational hazards, and risk-mitigation strategies; how to signal residual risks for the project construction and occupation stages.

Construction Strategy – to include site logistics, site set-up, transportation to and from site, and waste-reduction targets; it might also include criteria on selection of materials, specifications and characteristics, attitudes to technological innovation, and offsite manufacture.

Handover Strategy – this will set out criteria for commissioning all building systems, specific building-systems training for client maintenance staff, the frequency and protocols for snagging meetings, and zero-defects targets if applicable; it might include client demonstrations and handover of documentation, keys, etc.

Project Strategies reviews and updates

There is a clear value in developing Project Strategies throughout the two stages covered in this guide. It is beneficial to establish a plan for these strategies as early as possible in the project (ideally at Stage 1, as part of the Project Execution Plan and Initial Project Brief) and to maintain them by review at each Plan of Work stage. Project Strategies essentially contain the 'ingredients' of the building project, describing what the design team need to achieve in the performance of the building and acting as a set of guiding principles as the design and project teams decide between one priority and another during the course of the design project. A table of the essential Project Strategies and their review cycles across the first three Plan of Work 2013 stages is set out here:

STAGE 1 PREPARATION AND BRIEF	STAGE 2 CONCEPT DESIGN	STAGE 3 DEVELOPED DESIGN
Communications Strategy	Review and update if necessary	Review and update
Technology Strategy	Review and update if necessary	Review and update
Project Objectives	Construction Strategy	Review and update
	Health and Safety Strategy	Review and update
Procurement strategy	Review and update	Review and update
Design Responsibility Matrix	Review and update if necessary	Review and update
Sustainability Aspirations	Sustainability Strategy	Review
Project Outcomes	Maintenance and Operational Strategy	Review and update
Handover Strategy	Review and update if necessary	Review

Table 1.1
Essential Project Strategies and their review cycles across Stages 1, 2 and 3.

MANAGING THE DESIGN PROCESS
through Stages 2 and 3

All projects need a management structure to control their direction and progress. During the Plan of Work Stages 2 and 3, design leadership is a key mechanism for managing the required design processes. There is, however, an important distinction to be made between lead designer, who should be appointed during Stage 1, and other management roles in the construction business. While projects may have project managers, construction managers and BIM (Building Information Modelling) managers – all playing important roles – the lead designer has to be a person who understands the design process and can both construct and manage a credible Design Programme for the design team whilst also articulating design issues and process to the client and wider project team. Subject to these criteria design leadership can be awarded to any suitable member of the design team, but nominating the architect or an appropriate senior member of the design team on their appointment at Stage 1 is likely to be the most successful solution during the two subsequent stages.

DESIGN LEADERSHIP AND DESIGN TEAM WORKING

It is vital to the successful running of any project that the design team involved feel comfortable working together. Using opportunities to allow design team members to demonstrate their skill and worth to the process will help others to understand how they can contribute. Establishing clear design leadership at the beginning of a project, setting goals for team members and involving the team in the design-review cycle and a decision-appraisal process, will help create a respectful and fertile team-working environment.

Design leadership has common ground with other forms of leadership. The lead designer has to create conducive conditions for collective and collaborative working in circumstances that recognise all the design team members' value to the process. The lead designer must also inspire in design team members the confidence to contribute to that team's best work. How can a design team member introduce innovation to the process? When is it acceptable to introduce new thinking? How can design team members be encouraged to think beyond ordinary solutions when appropriate?

It is worth noting briefly that the dynamic of the design team can often be affected by who is employing each design team member and whether the lead designer role is carried out by a non-designer. Direct appointments by the client have the advantage of all design team members producing their best work to enhance their reputation and standing with that client. A lead designer-led team with other design team members employed by them can behave with different priorities. The lead designer has to find the right way to inspire the design team in either of these scenarios. If a non-designer consultant acts in this team coordination role, it may create a bias that the design team will have to work hard to overcome.

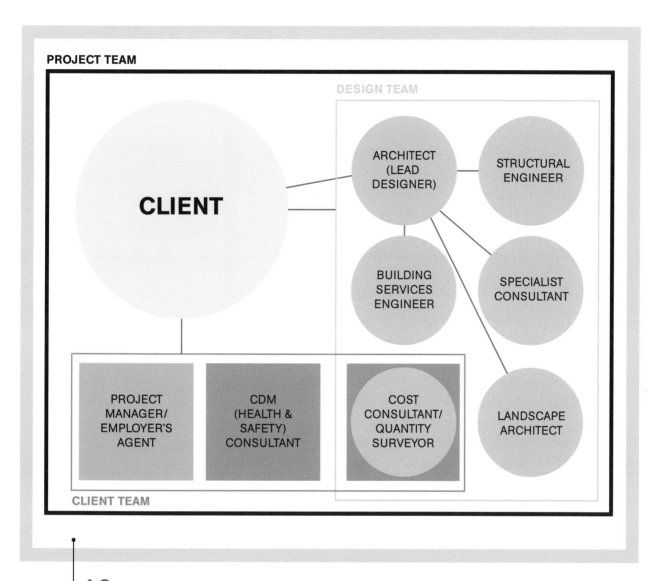

1.3
A team network diagram.

> **LEAD DESIGNER**
>
> A lead designer is a practice, or an individual, who can keep on top of Project Strategy reviews; articulate and drive the Design Programme; and maintain, throughout the development of the design, a clear sight of the project vision. During the two stages discussed in this guide, the nature of the lead designer will help to ascertain the quality of the Information Exchanges. It is important to distinguish the lead designer from the project manager or other management figures within the project team. These should be complementary roles, working together to ensure the smooth progress of the project and with clearly defined functions that relate to the key support tasks at each stage of the Plan of Work.

The lead designer during Plan of Work Stages 2 and 3 will often, but not exclusively, be the architect. For mid-scale to larger projects, it is worth considering this role being carried out by an individual other than the 'designer'. In other words, the architectural might have a lead designer and project architect alongside them in a team. This helps to distinguish between the job of encouraging all team members in the performance of their roles and coordinating the design team's efforts across all the Project Execution Plan criteria, and the production of the information necessary for that discipline's contribution to the design. In the context of Stages 2 and 3, this approach means that there is a significant design-orientated voice that can influence discussions amongst the project team around other task bar areas of the Plan of Work – for example, Procurement and Programme. These areas can have substantial impacts on the design quality of the building project, with the risk of curtailed design time and a decision to offload cost and design risk to a contractor early in the process leaving material and equipment specification choices at the mercy of lowest-price criteria.

How the task bars relate to Stages 2 and 3

In the RIBA Plan of Work 2013, a combination of eight task bars provides the mechanism for checking and assessing progress at each work stage. A series of core objectives is central to the progress of the project, complemented by variable support tasks that will change with the nature and complexity of different projects. The table opposite briefly sets out each task bar under Stages 2 and 3.

**TASK BARS FOR STAGES 2 AND 3
(BASED ON THE TABLE INCLUDED
IN THE PLAN OF WORK 2013 OVERVIEW)**

	STAGE 2	STAGE 3
TASK BAR 1 CORE OBJECTIVES	Prepare Concept Design incorporating necessary care design team disciplines. Establish Project Strategies and agree Final Project Brief.	Prepare Developed Design with coordinated design input from whole design team. Develop Cost Information and Project Strategies relevant to the stage.
TASK BAR 2 PROCUREMENT	Review procurement strategy for Stage 2 specific tasks that may lead to a decision on procurement route.	Review procurement strategy for Stage 3 specific tasks.
TASK BAR 3 PROGRAMME	Project Programme reviewed. Specific Stage 2 Design Programme used to log design meetings, workshops and events.	Project Programme reviewed. Specific Stage 3 Design Programme used to log design meetings, workshops and events.
TASK BAR 4 (TOWN) PLANNING	Certain projects may anticipate submitting a planning application at the end of Stage 2. Most projects will engage in pre-application discussions with the planning authority.	Most projects will submit a planning application during, or at the end of, Stage 3. Validation requirements for this application can form a significant part of the Stage 3 design team output.
TASK BAR 5 SUGGESTED KEY SUPPORT TASKS	Establish Project Strategies for key subjects – Sustainability, Maintenance and Operation, Health and Safety, Construction and Handover. Consider research opportunities. Review and update Project Execution Plan.	Review and update Project Strategies from Stage 2. Undertake third party consultations. Review and update Project Execution Plan.
TASK BAR 6 SUSTAINABILITY CHECKPOINT	Sustainability pre-assessment carried out, initial Building Regulation model, describe target environmental conditions and consider resilience to climate-change factors.	Sustainability assessment, interim 'Part L' model, review design to minimise resource use and waste.
TASK BAR 7 INFORMATION EXCHANGES	Concept Design with outline structural and services design. Established Project Strategies and outline Cost Information and Final Project Brief.	Coordinated Developed Design with appropriate Cost Information and Project Programme. Accompanied by updated Project Strategies.
TASK BAR 8 UK GOVERNMENT INFORMATION EXCHANGES	As required.	As required.

Table 1.2
Task bars for Stages 2 and 3.

It can be seen from the above table that the task bars can be used to introduce control mechanisms for balancing the project criteria of quality, cost and time. The first and second task bars, Core Objectives and Procurement, have a series of strategies that will control and monitor the quality of the project. These include Project Objectives, Quality Objectives and Project Outcomes. The Project Budget, also part of the Core Objectives task bar, will have been discussed in outline during the briefing period; during Stages 2 and 3, this element will inform the choices taken by the design team. For Procurement, Task Bar 2 sets the context for these choices with the priorities of cost and time set against design quality, making early choices about procurement vital to project success. It is often during these stages that good design procedures and conviction about the quality of a design bring about recognisably successful projects, as opposed to those that have let cost dictate the design.

RESOURCING PROJECTS

As this guide is intended to consider methods and techniques for undertaking a comprehensive use of the RIBA Plan of Work 2013 during Stage 2 Concept Design and Stage 3 Developed Design, it does not contain a detailed discussion of the project fees and resource allocation for a project.

For a wider discussion about these subjects, refer to A Client's Guide to Engaging an Architect (2013 edition) and Handbook of Practice Management (9th edition) – both from RIBA Publishing.

In the context of this guide, it is important to include mention of a Schedule of Services, in order to promote a clear understanding of what the design team members are employed to undertake. In the Plan of Work, this is accompanied by a Design Responsibility Matrix that identifies areas of responsibility for the design team, removing ambiguity over who is leading each element of the design. It is worth noting, however, that the RIBA Plan of Work 2013 can be of great assistance in illustrating to a client, as part of the appointment process, what the commission will involve and how its fee component relates to this process as a resource. This exercise must occur during Stage 1, making it more important during Stages 2 and 3 that the relationship between fee income and resource allocation is clear and forms part of the Design Programme for the project.

WHAT SHOULD WE EXPECT AT THE END OF STAGE 1?

For Plan of Work Stage 2 Concept Design to commence at all, there needs to be a completed Stage 1 Preparation and Brief for the project. Many architects acknowledge that it is their skill in assembling, understanding and working with project constraints that ignites their creativity, and a well-researched Plan of Work Stage 1 will provide a comprehensive set of criteria informing the design team processes at Stage 2 as the Concept Design starts to emerge. Later, at Stage 3 Developed Design, a well-researched and executed Stage 1 and Stage 2 robustly underpin the success of the emerging building design. As was outlined earlier in this chapter, the RIBA Plan of Work 2013 is able to be constructed within each stage using the task bar activities, but it uses the Information Exchange from the previous stage to develop design information through the current stage and to form a new Information Exchange point upon completion of the latter.

Information Exchanges

The Stage 1 Information Exchange will reflect the scale and complexity of the project. The scenarios that weave through the Plan of Work guides will illustrate to some extent the variation in information type and quantity at this stage. It should be noted, however, that irrespective of scale or complexity, every project should include answers to some simple questions at the Stage 1 Information Exchange:

- Who are the key members of the project team, and who is going to act as lead designer? A Project Roles Table should provide the answers to these questions.
- What is the overall time available for the Project Programme, and when are key milestones anticipated? The Project Programme will be reviewed during Stage 1, and it should be clear about the timetable and milestones to be achieved in Stages 2 and 3.
- What is the procurement strategy and what method will be used? When is a choice of procurement route anticipated? The thinking in this area may form part of the Initial Project Brief, with the expectation that it will be tackled during Stage 2.

- What is included in the Project Budget, and how will it be controlled? Cost Information is always going to impact on the nature and scope of Stages 2 and 3, and clear data on the Project Budget is critical.

A well-considered and comprehensive Stage 1 Information Exchange will include as the key document the Initial Project Brief, setting out the client's ambitions and requirements for the project. Alongside the brief will be Stage 1 Project Strategies – for instance, the Technology Strategy, Communications Strategy and Handover Strategy. The Project Execution Plan will encompass the agreed criteria for delivering the project, and will refer to the Project Strategies appropriate to that delivery.

Further details of Stage 1 Information Exchanges can be found in the first RIBA Stage Guide in this series, Briefing: A Practical Guide to the RIBA Plan of Work 2013 Stages 7, 0 and 1 by Hilary Satchwell and Paul Fletcher, and also in Information Exchanges: RIBA Plan of Work 2013 Guide by Richard Fairhead.

The excitement of winning a commission in these early stages is rarely dulled by the fact that the process followed on the current project will be very similar to that on the last one. It could be argued that the closer it follows the linear framework described and illustrated by the guiding document of the RIBA Plan of Work 2013, the better the prospects are for a successful outcome. This is in no small part due to the discipline of gathering all stage information into a presentable, searchable and understandable package, known as the Information Exchange, which carries forward the full intent and knowledge of the preceding stage.

The design 'answer' itself will, of course, be different each time, and will form part of the diverse and creative responses from a complex and evolving built-environment sector. The purpose of providing the RIBA Plan of Work framework has, since its introduction in the 1960s, always been to promote a good robust methodology underpinning those creative processes, which helps design teams to deliver consistent levels of service as well as good design.

WHAT SHOULD WE EXPECT AT THE END OF
Stage 3, to allow Stage 4 to start?

This subject is covered in detail in Chapter 3 of this guide, but, briefly, the Information Exchange at the completion of Stage 3 will represent a fully coordinated design that has been tested and costed to provide the client with the confidence to sign off the design, submit a planning application and proceed to Plan of Work Stage 4 Technical Design. The Information Exchange as a minimum will consist of:

- Design information from all design team members.
- The procurement strategy, indicating decisions taken during Stage 3.
- A Design Programme coordinated with the Project Programme, showing the proposed sequence and detail required during Stage 4 Technical Design.
- The Project Budget.
- The planning application; planning permission may be a prerequisite to starting Stage 4.
- Health and safety implications, which will have been considered and mitigated where possible by the design team, and a residual risk register.
- All Project Strategy documents; these key strategies will cover Sustainability, Maintenance and Operation, Handover, Construction and Health and Safety.
- The Project Execution Plan.

CHAPTER 01

SUMMARY

The Concept Design (Stage 2) is the expression of a core idea, around which the constituent parts of a project can be based. The design team produce a strategic design that meets the Initial Project Brief before advancing to Stage 3. The Developed Design (Stage 3) in essence sets out a series of further tests to validate the Concept Design, and normally concludes with a planning application submission. Project Strategies run through both stages, being reviewed from previous stages or established to inform an emerging element of the project. These strategies give the design process purpose and shape, and can help inform and steer the project team to a successful Project Outcome. Critical to this successful project is good design leadership that promotes, inspires and manages the design team to produce their best work, presenting a coordinated, costed and programmed building project to the client.

SCENARIO SUMMARIES

WHAT HAS HAPPENED TO OUR PROJECTS BY THE END OF STAGE 1?

A Small residential extension for a growing family

The architect has held a small number of meetings with the family and prepared an Initial Project Briefing document, a Schedule of Services and a fee proposal, which have been accepted. A discussion has taken place about what other design team members will be necessary to progress Stages 2, 3 and 4 of the project, and then also any specialist subcontractors that may need to undertake design work at Stage 4. These services have been captured in a Design Responsibility Matrix.

To complete the Stage 1 process, an outline Project Programme and Project Budget have been prepared. The architect has advised the client that a traditional procurement route will be the most appropriate for this type of building, and that they should consider how they might seek recommendations for builders.

B Development of five new homes for a small residential developer

The architect has prepared an Initial Project Brief for the developer, along with a Schedule of Services and a fee proposal. A Feasibility Study has established a capacity for the site using the developer's standard house type. During a discussion about the Project Programme, the developer has indicated that they would like to obtain planning permission at the completion of Stage 2 and that a traditional contract is their preferred procurement route at this stage. The architect has set out how these factors will affect the timing of events in the Project Programme, together with the Project Objectives and Outcomes, what other design team members need to be involved during Stages 2 and 3, and how the early planning submission may affect the design team's ability to fix the Project Budget before the completion of the Stage 3 Developed Design information. These points are presented as a project Risk Assessment.

STARTING STAGES 2 AND 3

Refurbishment of a teaching and support building for a university

New central library for a small unitary authority

New headquarters office for high-tech internet-based company

The university have prepared an Initial Project Brief using an internal project manager. The design team is being appointed as a separate entity, and the completion of Stage 1 has involved a review of the project brief; a Feasibility Study on the options for the refurbishment of the existing building, to be able to establish capacity; and an Initial Project Budget. The Project Programme has identified that site investigative and existing utilities survey work will need to be undertaken during Stage 2, so that the scheme is based on known information for the scope of the project and its cost and delivery period are supported at the intended planning submission at Stage 3.

The university's procurement strategy is to be adopted by the design team, and it anticipates a single-stage design and build contract with the design team novated to the successful contractor. A health and safety risk workshop has been requested by the client's directly appointed health and safety advisor, and this is to take place early in Stage 2.

A Feasibility Study has been completed, demonstrating the most efficient use of one of the council's town-centre sites. The design team appointed to carry out the Feasibility Study have completed that contract, and will have to take part in a competitive pre-qualification questionnaire (PQQ)/interview process before the commencement of Stage 2 if they wish to continue to work on the scheme.

Site investigation work was undertaken to assist in assessing site capacity, an ecology survey was commissioned and a BREEAM (Building Research Establishment Environmental Assessment Method) pre-assessment was undertaken and included a benchmarking exercise against similar UK-based projects.

The Feasibility Study reviews the comprehensive spatial brief included in the Initial Project Brief against the site. An outline Project Programme and a Project Budget that sets the cost parameters for the project and includes all anticipated construction costs, client direct costs and a relocation-and-relaunch budget for the library service have also been prepared. This information forms the Initial Project Brief, which is used to consult users and other libraries and has formed the basis of the OJEU (Official Journal of the European Union) PQQ design team procurement process.

The client has high expectations for this eco project, and is very keen to achieve a completion on the new headquarters as soon as possible but without compromising quality. To that end, the design team have all been appointed directly at the earliest possible stage after a competitive interview based on previous portfolios and references. The briefing period has been overlaid with a wide range of Site Information in order to gain the best possible information going into Stage 2, and it is expected that this information will help in pre-application discussions on planning being held with the local authority.

A short exercise on procurement and programme options has placed an emphasis on the speed of the process, and the procurement strategy compares the relative merits and drawbacks of various forms of procurement. The scoring matrix suggests that management contracting would provide the most appropriate form of procurement and the project team proceed on this basis.

25

CHAPTER 02

STAGE 2
CONCEPT DESIGN

RIBA Plan of Work 2013

Stage 2

Concept Design

Task Bar	Tasks
Core Objectives	Prepare **Concept Design**, including outline proposals for structural design, building services systems, outline specifications and preliminary **Cost Information** along with relevant **Project Strategies** in accordance with **Design Programme**. Agree alterations to brief and issue **Final Project Brief**.
Procurement Variable task bar	*The Procurement activities during this stage will depend on the procurement route determined during Stage 1.*
Programme Variable task bar	Review **Project Programme**.
(Town) Planning Variable task bar	*The RIBA Plan of Work 2013 enables planning applications to be submitted at the end of Stage 2. However, this is not the anticipated norm, but rather an option to be exercised only in response to a specific client's needs and with due regard to the associated risks.*
Suggested Key Support Tasks	Prepare **Sustainability Strategy**, **Maintenance** and **Operational Strategy** and review **Handover Strategy** and **Risk Assessments**. Undertake third party consultations as required and any **Research and Development** aspects. Review and update **Project Execution Plan**. Consider **Construction Strategy**, including offsite fabrication, and develop **Health and Safety Strategy**. *During this stage a number of strategies that complement the design are prepared. These strategies consider post-occupancy and operational issues along with the consideration of buildability. Third party consultations are also essential.*
Sustainability Checkpoints	• *Confirm that formal sustainability pre-assessment and identification of key areas of design focus have been undertaken and that any deviation from the **Sustainability Aspirations** has been reported and agreed.* • *Has the initial Building Regulations Part L assessment been carried out?* • *Have 'plain English' descriptions of internal environmental conditions and seasonal control strategies and systems been prepared?* • *Has the environmental impact of key materials and the **Construction Strategy** been checked?* • *Has resilience to future changes in climate been considered?*
Information Exchanges (at stage completion)	**Concept Design** including outline structural and building services design, associated **Project Strategies**, preliminary **Cost Information** and **Final Project Brief**.
UK Government Information Exchanges	Required.

CHAPTER 02

OVERVIEW

This chapter looks at what Concept Design is, and how the design team makes progress towards it during Stage 2. It examines how the design process, using the Plan of Work, shapes the quality of a project and the information that supports it at the completion of the stage. It explores how Project Strategies can be used along with other Plan of Work tools to build an exemplary client service, and explains what outputs are expected from Stage 2.

APPROACHING THE CONCEPT DESIGN

To many architects, Concept Design is 'the exciting bit' of the process of architecture. It might remind them of when they were students, thrashing around ideas with peers and tutors in order to pick up the essence of not only the practical outputs of the project in hand but also:

- Something of the ethos of the client.
- The cultural purpose of the building.
- The criteria that add to the 'narrative' of a designer's portfolio.

To other members of the project team, the terminology surrounding Concept Design stage has historically been within the ownership of the architect member of the design team, even though their own disciplines use the term to mean something similar – namely, how their own contributions contribute to the emerging design principles.

Recent best practice and the collaborative methods of working promoted by the RIBA Plan of Work 2013 firmly move the activity behind the Concept Design stage into the ownership of the wider project team, and set it in the broader context of how the project will unfold across future stages. This chapter helps to set out how all project team members might expect to take part in and contribute to this stage.

In the setting of the design team, the Concept Design always ranks high in excitement due principally to the potential at this stage to come up with a successful design which exceeds expectations and create a reputation-enhancing project. This process can appear artistic, magical and often unforeseen by the client. It might seem that this is the almost exclusive territory of the architect, putting together a mixture of historical and cultural references about the built environment with imaginative, aesthetic and technical skill, creating ideas that carry more than the sum of their functional and imaginative parts.

STAGE 2
CONCEPT DESIGN

Yet behind every Concept Design there has to be an organised and methodical examination of the facts and constraints; an incisive knowledge and inspection of the Initial Project Brief; and a strong, controlled sense of the essential components of any project – quality, time and cost. This level of investigation cannot be the exclusive preserve of one design team member. Indeed, steering a successful project through the early stages of development requires rigorous and complete input from each design team member in order to build confidence in concluding that stage with a strong robust design that has been tested by all. The lead designer should be appointed to steer the design team, ensuring that the developing design meets the Project Objectives by using various tools, including Project Strategies as set out during Stage 1 and reviewed during each Plan of Work stage. As a general rule, early and detailed considerations will pay dividends later in the process.

WHAT IS STAGE 2 CONCEPT DESIGN?

Plan of Work Stage 2 is the point in a project at which the design team start to prepare a response to the Initial Project Brief from Stage 1. The design team will review the Project Objectives, Quality Objectives and Project Outcomes set up for the project during Stage 1, along with other prepared information. They use this information and work towards a Concept Design for the project for consideration by the client. The key characteristics of the Stage 2 Information Exchange are that the client will obtain a clear understanding of how the building will perform, what it will look like, when it can be delivered and how much it might cost. These factors are addressed in a series of Project Strategies, which consider the specific and detailed aspects of the project and set out what needs to be considered in further detail during Stage 3. Project Strategies are discussed in more detail later in this chapter, but first some questions arising within the Core Objectives task bar are considered.

PROJECT TEAM AND DESIGN TEAM

Throughout this guide, the terms 'project team' and 'design team' are used to identify specific groupings of people involved in project delivery. For clarity about what these terms cover, a definition is included here:

Project Team – includes the project lead, who may be the client, and any agent that they may appoint to act as manager or expert in their interests with regard to the project, financial or funding advice, legal advice and the design team.

Design Team – includes the lead designer who is typically the architect but who may also come from one of the following disciplines: building services engineer, structural engineer or cost consultant. Depending on the complexity of the project, the design team may also include additional specialist consultants including: a conservation architect, planning consultant, ecologist, acoustician, fire engineer, facade engineer, or transportation and highways engineer.

DESIGN
A PRACTICAL GUIDE TO RIBA PLAN OF WORK 2013
STAGES 2 AND 3

WHAT DOES THE CLIENT KNOW?

There are many instances in which the client and design team will have previous experience working with each other, and this might inform how they choose to reach the beginning of Stage 2. In many other situations, the architect and other members of the project team will be meeting and discussing ideas for the first time. In these cases, an understanding of how much the client knows about the process and ensuring that the client understands what to expect from the design team and when is vitally important.

Further discussion on the range of client types can be referred to in the Handbook of Practice Management (9th edition) from RIBA Publishing.

During the early part of Stage 2, managing client expectations might utilise the following prompts:

- Use the Stage 1 Information Exchange to set up design parameters relevant to Stage 2.
- Be realistic from the beginning of the stage about what the project is going to deliver by the end of it, given what is known about the brief, budget and programme.
- Illustrate what the client can expect from their project by reference to precedent projects of similar scale, type and cost.
- Use the above parameters to set creative targets for the design team during Stage 2.

STAGE 2
CONCEPT DESIGN

It will be of significant benefit to the project for the whole project team to have the opportunity to gain a good understanding of the Initial Project Brief. Points worth considering in order to achieve this are:

- Develop commonly held Project Objectives early on in order to provide clarity of purpose.
- Acknowledge the role of all project and design team members in providing information for the Concept Design stage.
- Review the post-occupancy criteria set up in Stage 1 early in this design stage.
- Have a workable methodology for quickly checking cost and programme parameters as the Concept Design evolves.
- Work through a methodical and robust process towards Concept Design: this is enjoyable and rewarding, and can cement project team relationships for the whole project ahead.

HOW DOES THE PROJECT FIT INTO the design team's practice?

Irrespective of the size, scope or complexity of the project, it is useful to be able to assess during this early stage what potential the project has, as a whole or in specific aspects. This contributes to the collective knowledge base of the practice as well as individual design team members – for example:

- Are there new processes that need to be carefully negotiated?
- Is there scope for the introduction of new construction techniques, new technologies or materials? What research is required in order to establish what these should be?
- Do any of these considerations form part of the Initial Project Brief?
- Has the Design Responsibility Matrix produced at Plan of Work Stage 1 Preparation and Brief earmarked areas of the building's production information that individual design team members feel requires research during the early part of Stage 2?

The result of this enquiry may identify a particular research project that needs to be carried out in order to inform the Stage 2 process. It is important not to innovate for its own sake, but looking for opportunities to set up specific research projects, which may cross disciplines or be internal to architectural practice, helps to formalise a Research and Development strand to a practice – if this is not, in fact, already established. This is useful for improving the collective understanding of the practice and to develop a strand of work that could be productive in its own right. Expertise around published research can become a hallmark of quality, and of the rigour that a practice chooses to apply to the design process. This is rewarding for those involved, useful for both current and future clients, and excellent marketing by illustration of practice approach and methodology.

RESEARCH AND DEVELOPMENT: EXAMPLES OF FIELDS OF ENQUIRY

Research is required when experience and knowledge fall short of the requirements of a project. Formalising research can result in practice-wide, as opposed to simply individual, knowledge growth – for example:

- Research into a building typology new to a practice, when an established client wants something different from their usual architect.
- Types of offsite construction and impacts on cost and programme.
- Colour psychology.
- Wall-to-floor ratios to promote efficient envelope design.
- Best practice in daylighting for work positions.

Another useful exercise to factor into the processes being undertaken at this stage is a consideration of risks. This should form the basis of the Stage 2 Risk Assessment. If appropriate to the project's scope, a risk workshop involving all project team members might ask an iterative set of project questions, which could include:

- What risks lie within the project parameters themselves?
- Where, on the road ahead, are there obstacles or barriers to progress?
- If the above can be anticipated, how can they be avoided?
- What methodology can be employed to identify risks not yet foreseen?
- What lessons learnt from previous projects can be brought to bear on this one?
- What obstacle to a line of enquiry can be seen by one project team member but not by another?

Sharing the information being used to develop design ideas at this stage can avoid very significant problems later in the process. At this stage, there is time and good reason to adjust the concept direction in order to suit the collective knowledge of the project team; later on, there may be less time and little opportunity to properly address these issues, leading to a compromised design.

Where does design figure in the linear process towards a building?

It is no good any member of a project team, but particularly the design team, complaining about the lack of opportunity for design in the building and construction process. Every built project has a design component and, more often than not, design recurs throughout the initial, development and construction phases. How those moments are recognised, and who controls the series of decisions that are taken as a result of them, is critical to the role that design has in shaping and delivering a building. It is possible to influence design behaviour in the project team at this early stage by being clear and organised about how design decisions will be made and how coordination between design team members in accordance with the various Project Strategies is going to work. If the lead designer can generate acceptance of the fact that design is inherent in all decisions made within Stage 2, this will impact positively on the quality of the project being built and the user experience during occupation.

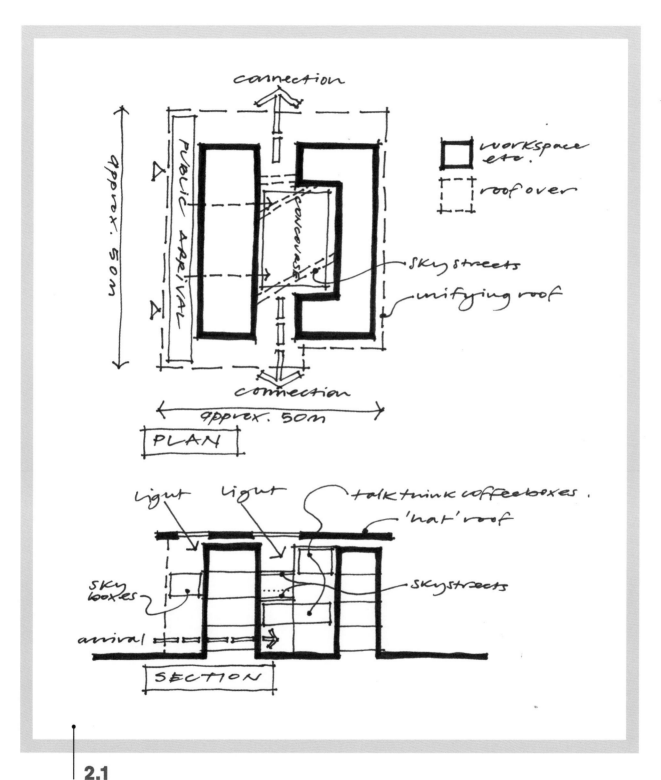

2.1
An illustration of a design concept.

DESIGN
A PRACTICAL GUIDE TO RIBA PLAN OF WORK 2013
STAGES 2 AND 3

What use are precedent projects for the type of project brief?

In responding to the Initial Project Brief at Stage 2, it is useful for the design team to assemble a series of precedent projects, or elements of projects, that illustrate previous building types: a so-called 'precedent study'. For domestic clients, this is likely to be pictorial; for professional clients, it might also include operating data on energy consumption, maintenance costs, etc. The precedent study can also act as inspiration for the design team on how they will test and respond to the Initial Project Brief. If this is done across the project team and the collected examples used during the early stage meetings and workshop events, then it will promote design excellence and heighten the design team's desire to produce a high-quality building.

2.2
A precedent sheet example.

STAGE 2
CONCEPT DESIGN

WHAT MAKES A GOOD PRECEDENT PROJECT?

A good precedent project helps in explaining to others involved in the design process where similar building types have been completed successfully. The success factor might be specific to part of the building or an element of operating data, but the precedent must clearly state what lessons are proposed to be learnt or avoided from illustrating it. The sort of overall project precedent illustrated might be:

- A building-type study with recommendations to visit, analyse the spatial plan or study the relationship of specific components of the brief.
- A demonstration of intended massing and response to a particular context.
- A façade-design treatment, including proportions of wall to openings.
- Sustainability credentials to act as a benchmark.

The sort of elemental precedent illustrated might be:

- Materials in use, explaining appearance and maintenance cycles, repairs and renewals.
- Doors, windows and wall openings.
- Glazing types for large sections of wall.

This process can be used to educate, illustrate and define ambition for the design process, determining the likes and dislikes of the client. It can also significantly inform the development of the Initial Project Brief into the Final Project Brief.

For community or group-based clients, a 'field-study trip' to look at similar building types, talk to peer clients and learn about their positive and negative observations of the process is of immense value. The trip provides plenty of informal time for the group to discuss the project, and how what they are seeing might affect how they want to tackle their own project. This exercise promotes a consensus on design-quality thresholds, how they will be discussed and set during the course of the Plan of Work design stages and how they can be evaluated at the end of the project.

> ### WHO GOES ON THE FIELD-STUDY TRIP?
>
> Field-study trips can quite easily appear to some like a luxury or a perk. They should be undertaken by the project team decision-makers in as small a group as feasible, and the itinerary should be full in order to ensure efficient use of the group's time together and to reduce the likelihood of the trip being seen negatively. The planning and timetable for the trip ought to include the following:
>
> - A comprehensive information pack for each member of the travelling group, to allow good preparation; this should include building plans and press cuttings where possible.
> - Clear reasons for each building visit in the context of the group's project, with clear research objectives to be answered.
> - Built-in time for discussion with host clients, users and designers if possible.
> - Built-in time for a discussion review of each visit within the group.
> - A designated photographer, to ensure a good photographic record to illustrate the trip to others in the client/project team.
> - A review workshop event after returning, at which observations/findings and recommendations can be fed back to the project team.

Precedents are also very useful as an organising tool for conceptual aspects of the design and the work of the project team. Representing an informal way of expressing different levels of experience, previous projects that have achieved high levels of satisfaction in use or due to some other factor might demonstrate:

- Award-winning prestige.
- Excellent sustainability performance.
- Use of Modern Methods of Construction (MMC).
- Successful use of particular materials.

Precedent studies can be included in the Stage 2 report, and invariably help to elucidate the Concept Design narrative in the Design and Access Statement at planning stage.

AN EXAMPLE OF A FIELD-STUDY TRIP: DANCE CITY, NEWCASTLE UPON TYNE

Supported by Arts Council England Lottery Funding, Dance City, an existing organisation based in Newcastle upon Tyne, had appointed Malcolm Fraser Architects (MFA) through a competitive interview process to design a new building in a prominent location in the city. Part of the briefing process comprised a discussion about precedent projects known to the team, and it was decided that the client, the Arts Council technical advisor and the principal and project architects from MFA should undertake a field-study trip in order to inform the Concept Design process. The buildings visited over a four-day period, along with their principal reason for inclusion, were:

- Dance Base, Edinburgh – the architect's own award-winning precedent project.
- London Studio Centre, London – functional spaces for large numbers.
- Sadler's Wells, London – functional spaces for small numbers.
- Royal Ballet, London – premium dance studios, high-spec 'back of house'.
- Linbury Theatre, London – flexible performance space.
- The Place, London - the client's UK benchmark project, small performance space.
- Tanzhaus, Düsseldorf – the client's European benchmark project.
- Choreographisches Zentrum, Essen – a mid-scale performance space.
- Vooruit Centre, Gent – an arts centre with mid-scale performance space.

The information gathered far exceeded expectations, tangibly influencing the brief and the Concept Design thinking. The time spent discussing what had been seen and heard, and how this was influencing client thinking, proved invaluable and cemented a core team relationship that served the project well over its duration.

Who is in charge of overall design objectives?

On most mid- to large-scale projects, the client will have appointed a team member during Stage 1 as lead designer to provide leadership for the design team. The lead designer sets up the criteria for establishing the design process, and manages the design team in the assembly of documentation for the Information Exchange at the end of Stage 2. When the client is an organisation, it is worth them considering at the outset of Stage 2 appointing a 'project champion' with seniority in the client organisation who can complement the lead designer role and ensure that good design principles are maintained as the project progresses.

PROJECT CHAMPION

The project champion must be a senior, authoritative person from the client organisation who understands the role of good design in achieving a successful project for the client and is passionate in delivering it. They will be an advocate for good design to project stakeholders, and provide inspiration for the design team. Individual clients, like small developers and homeowners, should be encouraged to act as project champions themselves, and feel able to play an active role as project lead in the evolving design process – in particular, during Stage 2.

For the right projects, the project champion has a pivotal role to play between project stakeholders and the design team, whereby they act as the 'guardians' of the Project Objectives. Irrespective of the scale of a project, or whether its champion appears to be an obvious choice (as with a domestic project), it is worth identifying this role with written and circulated objectives that can be referred to and measured at important stages of the process. On any project, formalising these objectives and sharing 'ownership' of them helps to galvanise the project team in their efforts to achieve them.

What does designing 'look like'?

The process of designing has many characteristics, and different design teams will have many different ways of illustrating how progress is being made. This chapter encourages collaborative working, and that involves being able to show 'work in progress' as readily as finished work. Often, the discussions that emerge around rough and partially modelled buildings get to the essence of the design intent and can inform the iterative design process in a qualitative way that the presentation of finished-looking design information cannot.

There are many ways to demonstrate this process, but these are some prompts commonly used by architects:

- Design parameters/constraints drawings.
- Figure-ground plan – describes the footprints of buildings against spaces around them.
- Overlay hand-drawn sketches of parts of plans, elevations or sections.
- Bird's eye or axonometric views, to establish design hierarchy or contextual relationships.
- Block massing diagrams.
- Cardboard or wooden scale models.

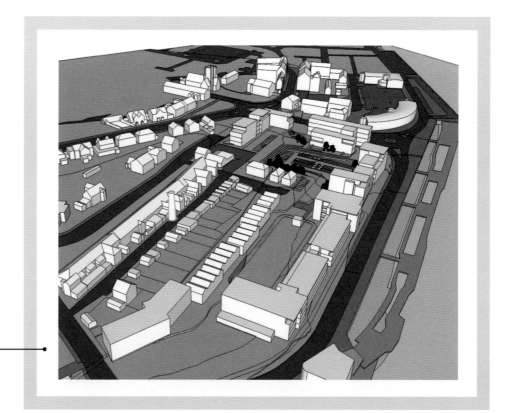

2.3 3D block model.

DESIGN
A PRACTICAL GUIDE TO RIBA PLAN OF WORK 2013
STAGES 2 AND 3

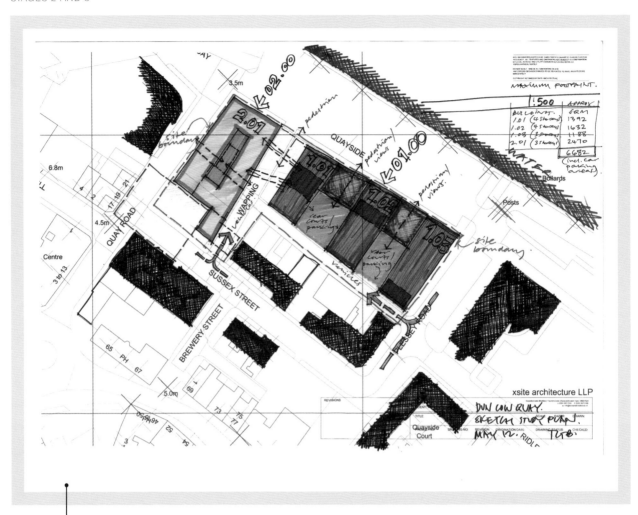

2.4
A site-capacity study.

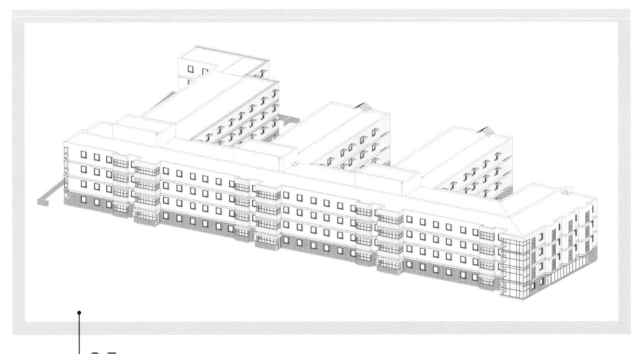

2.5
BIM image.

STAGE 2
CONCEPT DESIGN

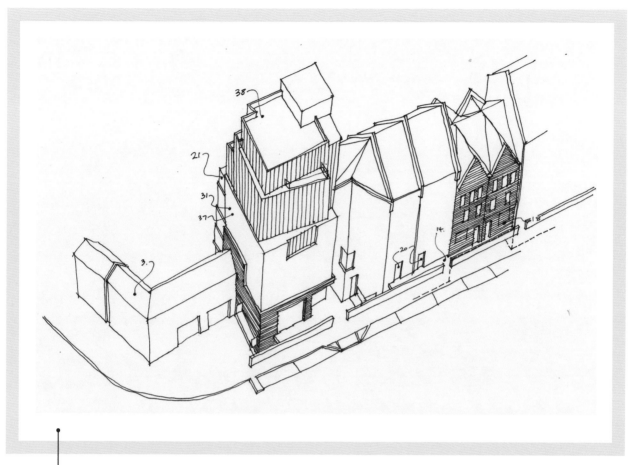

2.6
An emerging design sketch, exploring massing and materials.

2.7
A physical model:
North Shore prototype.

47

DESIGN
A PRACTICAL GUIDE TO RIBA PLAN OF WORK 2013
STAGES 2 AND 3

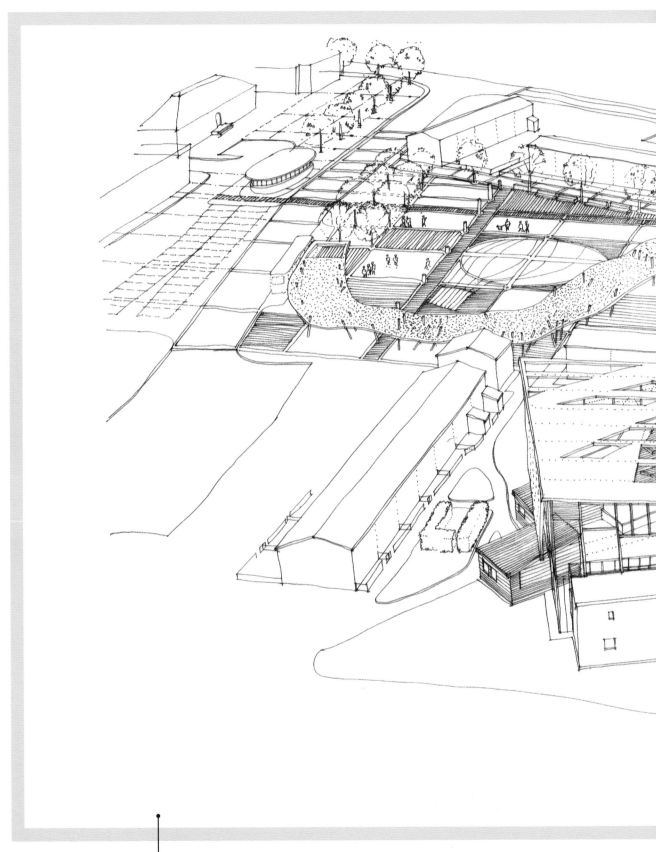

2.8
An aerial area sketch:
Byker Urban Renewal.

STAGE 2
CONCEPT DESIGN

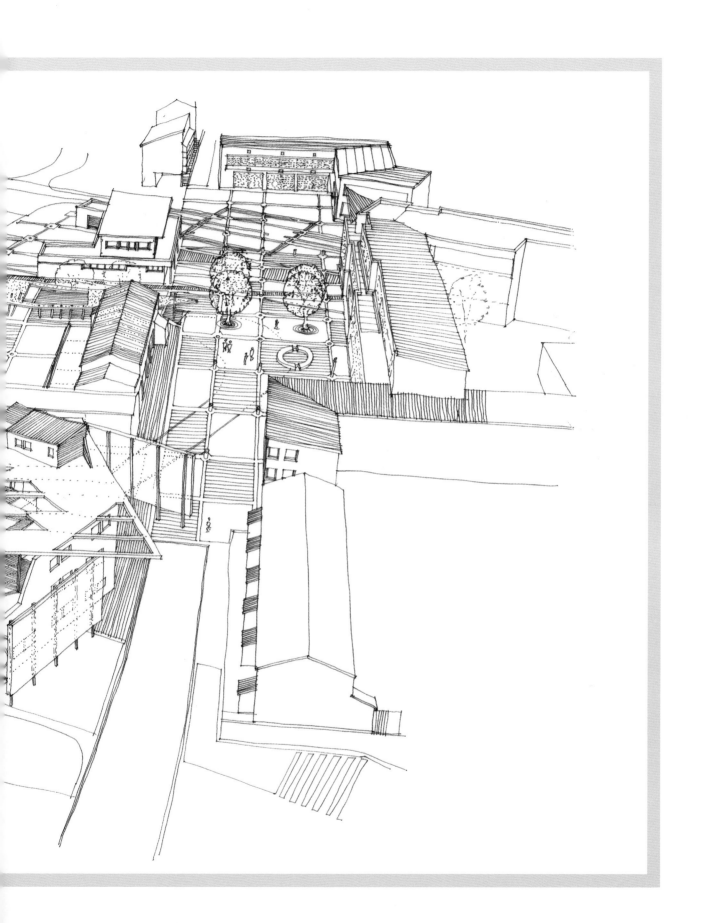

2.9
A sketch photomontage: valley view.

What is iterative design?

It is not very likely that the first drawn idea is going to meet universal approval. In fact, it is impractical to work in a collaborative environment in which iterative design processes are avoidable. The nature of iteration has to be planned and controlled, and should be seen as incremental steps towards the agreed design. There will be self-imposed design increments, information from design team members that informs progress, client commentary and external advice on criteria like design constraints or planning policy – all of which will drive iterations of the design process. The skill of the architect is to see clearly the purpose of each iteration, acknowledging that incremental improvements in design and the fulfilment of briefing criteria increase the qualitative outputs of Stage 2 and that this has long-term benefits for the project as it moves into future stages.

HOW TO DESIGN TO COST AT STAGE 2

Cost is a critical component in any project, and the lead designer and architect should ensure they understand cost criteria as well as the cost consultant does. The presentation of Cost Information during and at the completion of Stage 2 may take different forms depending on the chosen or expected procurement route, client needs and who is in the design team at this stage. (Some projects may not engage cost consultants until a later stage.) In all cases, however, clarity is a key objective, with the basis of cost calculations, inclusions and exclusions clearly identified.

With traditional types of procurement, cost information is firstly an estimate based on historical cost data. The more design information there is to work from, the more comprehensive the cost estimation can be. It is worth noting, however, that even in such circumstances it is only once tenders from the construction works are returned by contractors that a real cost for the project exists.

In other procurement scenarios, degrees of cost certainty can be requested by the client much earlier in the process, and this can have a significant effect on the design outcomes. For example, a cost cap that stretches the budget over too large a building will erode design quality. With that in mind, considering the design process as 'designing to cost' rather than 'costing the design' will help to steer the design team during Stage 2 to conceive of a Concept Design that delivers a high design 'tariff' at the same time as remaining affordable throughout future stages.

COMMON TERMS USED IN COST INFORMATION

End-of-stage cost models are often called:

- Outline cost plan – used very early on in the design process, and generally based on measured areas of an element of construction – for example, a square-metre rate for Gross Internal Floor Area (GIFA), hard landscape, soft landscape or highways external works.
- Cost plan – still using measured areas of an element of construction, but applied to more detailed parts of the building.
- Cost estimate – using costs of elements of work, and multiplying by occurrence – for example, number of doors or windows; area of roof; area of floor with covering A, area of floor with covering B.

Other terms used in cost models:

- Provisional sums – rounded numbers to cover items of cost that are not defined yet; in some forms of building contract, provisional sums can be carried into the contract sum.
- Contingency – a sum of money to cover unforeseen elements of work; often expressed as a percentage, which should decrease at each stage as more is known about the design.
- Design development contingency – a sum of money anticipating increased costs as the design becomes more detailed during Stage 3; often expressed as a percentage.
- Preliminaries – often expressed as a percentage, this figure covers the contractor's costs in carrying out the contract; for example, the site management costs, site welfare, scaffolding or plant hire.

The desire of the client will always be to have the best building they can afford, and behind that proposition lie a number of factors that should influence how the design team approach Stage 2 work – and particularly how they set up the conclusions of Stage 2 to be picked up at the beginning of Stage 3 Developed Design.

STAGE 2
CONCEPT DESIGN

AN EXAMPLE OF DESIGNING TO COST

If the project brief requires a 1,000 sq m building and the cost plan historic and indexed data says that this type of building costs £1,500 per sq m (all figures excluding VAT), then the construction cost estimate is £1.5m. If this figure is felt acceptable, then design develops towards tender. Let us say that the accepted tender return cost is £1.65m (£1,650 per sq m) for a number of reasons and the client is insistent on the contract sum being the budgeted £1.5m, then the design team has to find £150,000 of savings. This is equivalent to having 90% of the building you thought you were buying, or to savings from across the cost plan eroding the design intent considerably.

If, in the first instance, the £1.5m Project Budget had been tested against projects with contingency or inflationary factors considered more fully, then the design team would have been able to say to the client that they can afford a 900 sq m building or that the Project Budget would need to increase to accommodate the additional estimated costs.

When designing to cost, this approach of presenting costs with better contingency levels would be supplemented by having as many large elements of the design as possible tested in the marketplace with budget costs. For instance, the structural-steel package, the envelope package or the services-installation package could be tested in order to produce subcontract cost information to inform the cost planning at Stage 2. This process does not lend itself easily to all forms of procurement, but it can form part of a constructive dialogue on cost with both suppliers and design team members who can modify designs to include specialist commentary from the process.

One factor that often surfaces during Stage 2 is that some members of the design team seek to leave elements of design unresolved until the next stage. The temptation to do this may relate to time allocation in the Project Programme, fees assigned to resources during this period or an unwillingness on the part of one or two design team members to close down the options within their discipline until more information is available in subsequent stages. The Design Responsibility Matrix (DRM) reduces these ambiguities, making deliverables clear, and the lead designer needs to be on top of this situation from the outset and agree with the cost consultant how cost information is going to be dealt with during this stage.

TIPS AND TECHNIQUES FOR PRESENTING PROJECT COSTS

- Separate out costs associated with easily verified elements of the building (eg new-build superstructure) from elements that cannot be based, at this stage, on known or quantitative cost information.
- Define clearly parameters used for estimating ground works, abnormal conditions and unknown elements. Describe what work is required to help firm up these early cost estimates.
- Clients will always be tempted to chop down design development and contingency sums, and it is useful to have some element of contingency placed against elemental headings as well as the project cost.
- On some projects it is worth avoiding, where possible, a long list of exclusions, as these often represent some cost to the project – for example: furniture and fit-out costs, professional fees and acquisition costs. Make an allowance for as many of these elements as you can, being clear that they are direct costs to the client and not part of the construction budget. This gives clearer sight of what the construction budget actually is.

It is self-evident that at the end of Stage 2 the client would like to understand the relationship between the Concept Design and cost. In tandem with this there is, of course, a programme component and an understanding of the risks associated with the cost plan. It is also self-evident that the more investigative and proving work that can take place during Stage 2, the more robust and informative the cost plan is going to be. After that, the presentation of possible additional cost factors may be better expressed as a percentage probability of expenditure rather than provisional sums that cover unresolved design elements. This method better reflects the complexities of a project, and is more likely to inform good decision-making at the conclusion of Stage 2.

REVIEWING THE PROCUREMENT STRATEGY

Although it seems a long way off, choosing the appropriate procurement route for any project is a key factor in a successful and well-run process. As part of the Initial Project Brief and Stage 1 procurement strategy, the client will have outlined their order of priorities for the three principal components of any project – quality, time and cost. For example, the client might have a rigid, maximum Project Budget figure but not be concerned about the period of time in which the project is delivered. In that case, the client's question might be 'How much building can this much money buy while maintaining high design-quality standards and without restrictions on time?' Another question might be posed illustrating that the time factor is important, the cost critical but that quality is the third priority. This question might be 'How quickly and cheaply can you deliver this simple shelter?'

Reviewing the procurement strategy during Stage 2 will only take a short time, but is worth doing to remind the design team of how the design information that they will produce will eventually be used. Even if a final decision is not made at this stage, the review may influence the types of information or levels of detail for elements of design that the design team produce both at Stage 2 and later. The review should be the subject of a specific workshop or a design team meeting agenda item, so that enough time is dedicated to exploring the options of this very important part of the project delivery.

IMPACT OF PROCUREMENT ROUTES ON STAGE 2

- **Traditional** – A Concept Design emerges with clear levels of design information from the whole design team. The Information Exchange clearly sets out where Stage 3 is going to improve project and cost information, and often concentrates less on the measurable parts of the project as a result.
- **Design and build single-stage** – If a tender is to be released at the end of Stage 2 to find a contractor to lead the process from Stage 3, then Stage 2 information has to translate the design intent and costed design information into a set of employers' requirements that is measurable against the changes that will inevitably take place later in the process.
- **Design and build two-stage** – The parameters need to be tightened for entering into Stage 3, where the contractor sits with the design team and they develop the design and cost information alongside one another.
- **Management contract** – The level of separation in building elements will have a great bearing on the Information Exchange at Stage 2, and more attention than is usual at this stage will need to be given to the coordination of elements of the project. The early relevance of construction information for certain packages will also need considering.

A key component of information that will be driven by procurement decisions is the specification document for the project. In general, the earlier the client and design team wish to fix the cost and let a contract, the more detailed the specification needs to be for that stage. Specification is a way of describing the performance of a building component, or the actual product or equipment that is required to meet the desired function or quality. Specification safeguards quality by being as explicit as possible about what the contractor should build.

STAGE 2
CONCEPT DESIGN

REVIEWING THE PROJECT PROGRAMME

The Project Programme will have been generated during Plan of Work Stages 0 and 1. It is created at the outset of the project, so that periods of time can be allocated to sequential stages of the project. It is also flexible enough to be able to illustrate how stages that would normally be sequential may need to overlap because of procurement decisions, and how that is to be managed. The Project Programme is reviewed at every stage, but it is also an excellent management tool for the project. It will have important, distinct components during Stage 2. The Design Programme, which is prepared by the lead designer with inputs from other design team members, and the Construction Programme are both important in establishing timeframes for each stage. In many projects in which a tender process and the appointment of a contractor is not planned until Stage 4, the Construction Programme line in the Project Programme will be a single bar of an estimated duration. If the procurement route chosen means that a contractor has already joined the project team, then the Construction Programme can reflect the level of detail known or assumed at the stage and will help inform the order in which decisions might need to be made in order to be able to achieve the Project Programme dates.

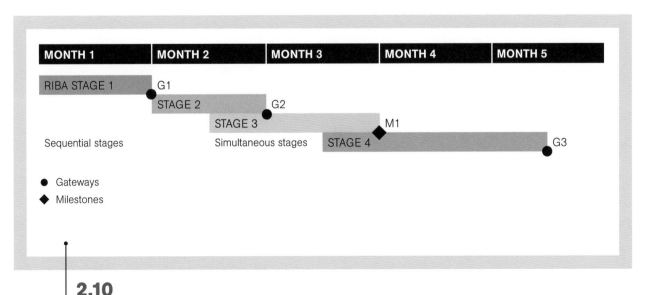

2.10
An illustration of a Project Programme extract.

The lead designer can use the Design Programme at this stage to ensure that Project Strategies are created or reviewed in a timely manner, in order to maintain progress. They can programme workshops, stakeholder meetings, external events, and third party and internal meetings. A determined rigour to complete this information not only informs the project in a task-orientated and clearly sequential way, but also coordinates all the Project Information into a timeline that is clear and can be easily referenced by all the project team members. Good information, delivered in a timely way, assembled to ensure a smooth transition to the next stage of the Plan of Work marks out the Design Programme as central to the success of the project. The Design Programme is often used as a 'dry' blunt tool when it should be active and alive with information transfers logged, new ideas shared, meetings and formal events noted. During the Concept Design stage, the programme records ideas and logs decisions in order to form a record of the stage for future reference if necessary.

How do ideas get organised?

Perhaps too often in architectural practice, Stage 2 activity will feel constrained by the period set for it within the Project Programme. Using a Design Programme to keep a clear framework of the order and timing of all design ideas and approaches will help to counteract this perception of time constraint. To assist with the organisation of this early Stage 2 thinking the Design Programme can log those ideas that predate an appointment, those that had average or good information on which to be based at the time they emerged and catalogue where ideas need further information to be fully tested against project criteria. It is necessary, as a core principle of the Plan of Work, to establish a Design Programme in the early part of Stage 2 – particularly on larger projects. Developing a Design Programme for a project offers the chance to illustrate how many discrete but connected activities there are in producing the Concept Design stage, and in what order they need to be so as to inform each other. This is useful as a visual tool as well, because it reflects a quantity of design work that progress can be checked off against.

STAGE 2
CONCEPT DESIGN

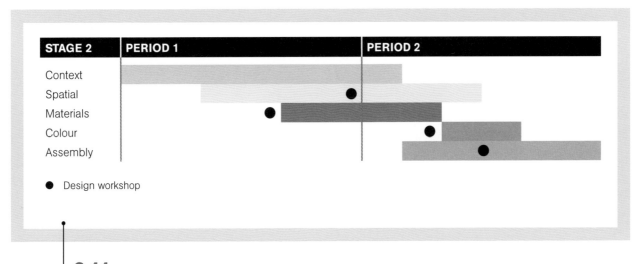

2.11
An extract from a Design Programme.

There is an innate, logical sequence in which Concept Design emerges. Most designers would recognise this as a process from large-scale thinking about the whole project, through stages of considering parts of the project to eventually addressing the relatively small detail. All architects know that the user experience of a finished building starts with material choices, textures and colours, and an appreciation of the detail in material junctions. The Design Programme can be organised to reflect this way of thinking if it is appropriate for a particular project. It may be that the entrance gateway of a project is the critical design statement, or that a particular method is required for refuse and recycling collection that will characterise the project layout. These idiosyncrasies need ideas for their resolution earlier in the process than the simple 'big to small' sequence allows, and the lead designer can influence this alternative sequence through their use of the Design Programme.

What should happen if the project is to be submitted for planning permission at Stage 2?

The Programme task bar (Task Bar 3) on the RIBA Plan of Work 2013 is designated as a Variable task bar, along with (Town) Planning (Task Bar 4) and Procurement (Task Bar 2). This underlines the fact that certain actions overlap or can occur at different stages. In general, it will be a reasonably infrequent occurrence that a client will decide to submit a project for planning permission at the end of Stage 2 rather than at the end of Stage 3.

This scenario would cover projects where the client wished to establish the principle of development on a particular site while limiting their committed expenditure. It is also reasonably common for projects to be submitted during Plan of Work Stage 2 because the client is interested in establishing a site value that is generated by a quantum of development rather than a 'market value'. While this situation often means a less robust design process during Stage 2 – and often no other design team members being involved, as the core aim is simply to establish a monetary value – it is quite possible, using previous experiences, to factor in design tolerances so that a credible planning application can be made. This approach has to be very carefully managed, and is not recommended for unfamiliar building types or reasonably complex buildings that are intended to be built.

DEVELOPING PLAN OF WORK
Project Strategies

Task Bar 5, Suggested Key Support Tasks, encourages the development of Project Strategies for all relevant aspects of the project. Each strategy can state clearly the ambitions of the project and how it is positioned in relation to industry norms. They can compare tried and tested strategies or those from other projects, and establish what criteria will be used to steer the project through to a successful construction stage. As examples, Stage 2 may include the development of strategies for:

- Sustainability – including energy efficiency and whole-life costing.
- Fire engineering – resistance criteria for elements of structure, means of escape and fire-engineering inputs required.
- Acoustic – dealing with noise sources external to the project, or generated by the project operation.
- Security – what levels of access control are required at the site boundary, the building entrance and in circulation routes around the building?
- Health and Safety – assessment, management and the recording of health and safety issues.
- Building façade design – design parameters for façades of a building.
- Spatial quality – design assessment, review and scoring.
- Construction – use of site access, site set-up, contractor welfare, etc.

It can be seen that some of these strategies are technical in nature and others are cultural, economic or design orientated. These strategies will help explain to a client how aspects of the design process are being considered in detail. They will form a key part of the reference point at the conclusion of the Stage 2 process, and will be useful as a coordination tool for the lead designer. They must have built into them the review process at future stages and ideally be formatted in a way that allows for updates to be made easily, so that progress can be mapped and a Feedback loop can help to share the project knowledge around the design team.

DESIGN
A PRACTICAL GUIDE TO RIBA PLAN OF WORK 2013
STAGES 2 AND 3

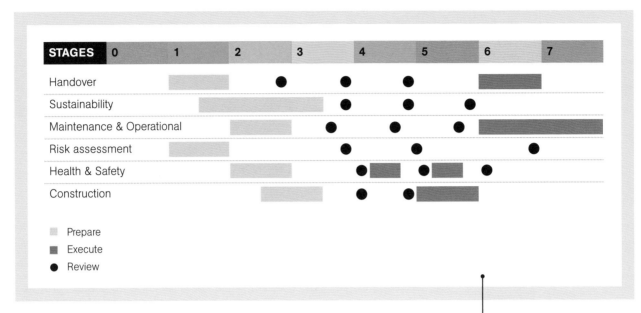

2.12
Project Strategies and their relationship with work strands within the Plan of Work stages.

Trackers for Project Strategies

Developing a methodology for registering and updating project factors that are generated internally to the design team or imposed externally are an important management tool for the lead designer. A task-tracker document managed by the lead designer is a particularly powerful way of communicating the status of the design, cost items or actions required to develop the Project Strategies. In future stages, different trackers can be generated for planning conditions or Building Control conditions and, eventually, Construction stage information. For a small project, this could be a single document that tracks tasks and progress across the whole Project Programme. For larger or more complex projects, it might be more manageable to have Plan of Work stage trackers or task bar trackers, eg Procurement actions against Task Bar 2 or planning conditions against Task Bar 4. However this information is recorded, it is crucial that the lead designer regularly circulates updated trackers to the project team for information and action.

A tracker document can also be utilised to map the progress of a particular Project Strategy that runs throughout the project – for example, the Sustainability Strategy. In aligning information from across the design team in one document, it becomes more easily understood how design decisions across the team impact on one another. The lead designer maintains ownership of these documents, and it should not be underestimated how much resources this task will require. The project benefits substantially from better and more widely available information, and project trackers can form a significant part of the Plan of Work stage Information Exchanges.

NO.	DESCRIPTION	STATUS	ACTION	BY	NOTES FOR STAGE 3 (PLANNING)	NOTES FOR STAGE 4 (PRODUCTION INFORMATION)	PROGRESS TO DATE
1	Technical check required on proposed glass sheet sizes		Meet product representative	AA	—	—	Call made 01.04.14
2	Footprint of building covers too much site		Agree percentage target	PC	Key r/ship with context		Client meeting necessary
3	Conflict in brief around visibility/privacy		Agree strategy	AA	Review strategy		Client meeting necessary
4	Review whole project materials palette		Complete materials review; contact manufacturers	AA	Fix palette and cost plan	Gather all technical literature before Stage start	Samples ordered, Outline Specification drafted
5	Issue Stage 2 updated Maintenance and Operational Strategy		Programme Design Workshop for this item	LD	Review strategy	Review strategy	Meeting set up with stakeholders
6	Carry out early Design Risk Assessment		Review and update before stage completion	PD	Review and programme Risk Workshop	Review and programme Risk Workshop	Issue no. 1 25.03.14

- ■ To start
- ■ In progress
- ■ Urgent action required
- ■ Outstanding queries
- ■ Completed

2.13
Example of Plan of Work Stage 2 quality strategy tracker.

WHAT IS THE PROCESS FOR REVIEW OF the Project Execution Plan?

The Project Execution Plan (PEP), when prepared correctly, is a key document that comes into being during Stage 1. It contains a collection of all the available information relating to the delivery protocols and processes that the project team will adhere to. It will have been prepared by the project lead perhaps in conjunction with the lead designer, and it makes clear expectations about how information and computing technology might be used in a Technology Strategy. This will include the standard formats for document exchanges between project team members, aligning elements of naming and numbering in order to enable universal understanding across the project team, and will establish the principles of collaborative working, formats for key Project Information and any bespoke processes for the project.

At the beginning of Stage 2, the PEP will be briefly reprised. This is particularly useful if there has been a programme break between Stages 1 and 2, as there are likely to be new members of the design team. This reprise can happen at a stage start-up meeting, electronically or via a specific workshop.

A further review of the PEP will be needed part-way through the Stage 2 period, so that an accurate update will be available as part of the Stage 2 Information Exchange. On large and complex projects, it may be pertinent to have the PEP review as a 'live' document on every design team meeting agenda. The review should include the whole project team, acting as a reminder of the project principles but providing the opportunity to catalogue and communicate all of the updated sections. The PEP must be kept up to date: it is one of a suite of project documents that are created at the beginning of projects, and not maintaining it can lead to a loss of focus, indiscipline in the production of Project Information and the consequent erosion of Project Objectives.

HOW CAN DESIGN WORKSHOPS
be used productively?

Stage 2 Concept Design processes will be improved by the introduction of design workshops at key points during the process. This approach invites the whole design team, sometimes with the client, to investigate collaboratively various approaches to design problems. An identified problem from one design team member benefits from the immediate input of each discipline around the table, and, even if a solution is not settled on immediately, the further investigation of that problem will be better informed than previously. The design workshop can avoid the 'how we did it last time' default position, and should be planned by the lead designer as an entirely different experience to a design team meeting or project meeting.

EXAMPLE OF CONTENTS OF A DESIGN WORKSHOP AS AGAINST A DESIGN TEAM MEETING

DESIGN WORKSHOP	DESIGN TEAM MEETING
Introduction to subject area for workshop – for example, building management and operation, public realm, or landscape or envelope design	Review minutes of last meeting
Review project brief	Actions from last meeting
Review precedent examples	Health and safety
Design team members explain the criteria for their scope of work relevant to the subject	Reports from all design team members
Working session with sketch ideas and scenario planning	Project Execution Plan – check, review and update
Summaries of progress made	Cost
Impact of Project Strategies	Programme
Agreed actions log	A.O.B.
Outputs: notes, drawings and diagrams, actions log	Outputs: minutes, actions log

Table 2.1
The contents of a design workshop and a design team meeting.

As part of the Stage 1 Information Exchange, all project team members will have received a copy of the Initial Project Brief, including Project Objectives, Quality Objectives, Project Outcomes, Sustainability Aspirations, Project Budget, Project Programme and any Feasibility Studies and Site Information. This is a great deal of information – especially for small projects – and understanding this information as a summary presented to a design workshop is key to aligning the design team's engagement with the aims of the project in hand.

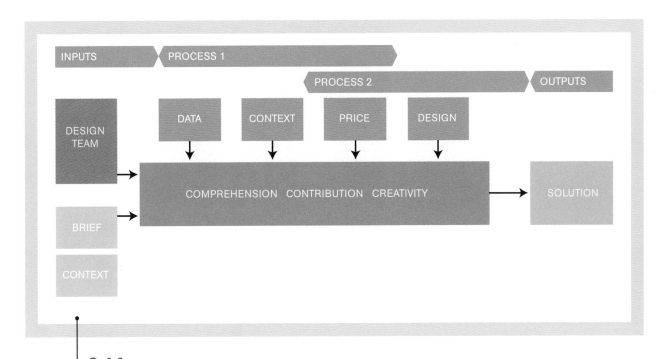

2.14
A design workshop process diagram.

The workshop might have quite a free-ranging agenda, so that lines of enquiry raised during the session can be considered properly and 'chased down' to a conclusion based on their relevance to the project. Some elements of the design might require a Research and Development component to maximise their beneficial impact on the project. While this type of engagement is healthy, and marks out the key difference between a design workshop and a design meeting, the facilitator must use their agenda and briefing materials to ensure that all attendees are able to make their relevant contributions. At the outset of the workshop, it is useful to set the tone for the session and, for instance, establish that 'all ideas are welcome and at this stage treated equally'. Another factor is to strike a balance between experience from a long career in the industry ('that doesn't work because…') and ambition born of a short career ('that detail tells the whole story of the building and we have to afford it…').

Before the end of the workshop session, there ought to be a short summary of the discussions that have just taken place and the key actions that arise from them noted for distribution. Using a tracker document as described previously is a good way of doing this, and acting as a reminder of the Project Objectives at the same time.

The energy that can be generated from an inclusive, positive design-workshop experience at the beginning of a project can carry huge benefits for how contributors perceive the project in their own workload, and ultimately how the project has the potential to enhance the reputations and portfolios of the design team involved.

> **DESIGN WORKSHOPS AND COLLABORATION**
>
> Pre-planning how these sessions are going to be organised, facilitated and noted, and structuring the lines of enquiry that design team members are encouraged to make, is critical to fostering joint ownership of the project's momentum and efficacy. Having an organised, shared platform from which to develop the project and to robustly test concepts against constraints and the brief increases the chances of the concept surviving future stages and forming the basis of clearly translated design intent as the project progresses towards Stage 5 Construction.
>
> Successful collaboration within the design team is best delivered through design-workshop events, and involves understanding the criteria for experimentation that exists in the work patterns of your design team colleagues. This feeds understanding of what the nature of redesign and change means to them, so that your own thinking is modified in relationship to the project as a whole – and not just your own disciplinary boundaries. It should be noted that simply sharing information is a poor substitute for the collaborative benefits of design workshops.
>
> Long enjoyed by multidisciplinary firms, it has recently become common for co-located working (or 'co-working') to be employed during the Concept Design stage, when periods of time spent in one office or studio can be quickly productive because the proximity of your co-working design team brings prompt answers or commentary that can progress your own design in the knowledge that it is safe to do so and without impacting on the efficacy of the design process.
>
> For example, a building services engineer based in Edinburgh, working with architects based in Newcastle upon Tyne on two projects, spends two whole days a week in the architects' office during Stage 2 of one project and Stage 3 of the other. This is concentrated time that allows significant progress during this co-located time and affords an insight into the working practices of each discipline.

SUSTAINABILITY CHECKPOINTS

Task Bar 6 provides for sustainability checkpoints at every Plan of Work stage. These checkpoints have developed from the 2011 publication Green Overlay to the RIBA Outline Plan of Work by Bill Gething (RIBA Publications). The sustainability credentials of any project are now central to all design processes. Whether the client has an ambition for an exemplar project or simply to meet current Building Regulations and relevant standards, achieving the best sustainability outcomes for projects can characterise both their process and their outcome. A sustainability checkpoint at Stage 2 should contain the following type of questions:

- Confirmation that a formal sustainability pre-assessment has taken place, and that key areas of design focus have been undertaken during the stage. Any areas that differ from the Sustainability Strategy should be highlighted.
- For the production of the initial 'Part L' model, has the design team considered a 'Fabric First' approach? (Fabric First is the strategy of prioritising an airtight, highly insulated envelope for the building before considering energy sources or heat or light emitting equipment.)
- Has the design team considered energy-efficient fittings and equipment?
- What renewable technologies might be appropriate for consideration in this project?
- Is it a high-tech control environment or a low-tech user-controlled project? Have 'Plain English' descriptions been used for internal environmental conditions, seasonal controls and control systems?
- Does the Construction Strategy refer to responsible material sourcing, recycling and environmental impacts?
- Are offsite prefabrication, recycled materials or local sourcing possible?
- Is a BREEAM, LEED (Leadership in Energy & Environmental Design) or similar environmental code or standard required?
- What is the design life of the building, and how will that be measured? Has resilience to future climate change been considered?
- How is the building to be operated? And by whom?

INFORMATION EXCHANGES: THE PLACE
of sketches, drawings and documents in the process

Task Bar 7 in the Plan of Work comprises the Information Exchanges at each stage completion, requiring sign-off from the client in order to proceed to the next stage. At the end of Stage 2, the Information Exchange could include the following:

- An updated Project Execution Plan, containing Project Objectives and Project Outcomes.
- A Final Project Brief that collates the amended briefing information from Stage 2 and updates the Initial Project Brief to reflect the Concept Design.
- Concept Design represented in agreed formats and media. These may include a building information model and/or spatial design drawings. Both would include outline structural and building-services information. Each design team member will produce information in agreed formats in order to describe the following:
    - ~ Architect – site context and building relationships; internal spatial arrangements; massing; heights; key strategic information related to layout, outline materials and finishes; specification; schedule of accommodation.
    - ~ Structural engineer – structural grid, structural zones required, substructure strategy, outline specification.
    - ~ Building services engineer – initial 'Part L' model; distribution strategy; sustainability statement, including energy consumption and renewables strategies; outline specification.
    - ~ Landscape architect – landscape strategy, spatial plan and outline specification, and, possibly, draft planting schedule.
- Project Strategies, including a procurement strategy, Sustainability Strategy and Construction Strategy.
- Project Programme and Stage 3 Design Programme.
- Cost Information, in an agreed format.

It is useful to bear in mind what the Information Exchange contents are going to be at the outset of the stage. It is most likely that the content will be predominantly digital in media, and some thought is necessary in order to ensure that transfers of this information can happen in a secure environment and in a timely manner.

The likely range of information at all Plan of Work stages is increasing all the time, and will very likely include access to Building Information Modelling (BIM) digital files as well as any hand-drawn material that is felt to be important to the whole narrative of the stage. The 'back of the envelope sketch' analogy is still relevant on occasion, and such drafts can be introduced by any member of the design team.

More than at any other stage, the draft copies, test prints, tracing-paper overlays and document versions and revisions produced by the iterative design process and the collaborative working of the design team leads to a significant output of material. Is it worth keeping all these layers of process, which may never be referred to again? A quality management regime may dictate the answer to this question, but what is certainly true is that this momentary information tells the story of the project and should be methodically archived for future recall – even if this means just being able to reproduce the 'eureka moment' sketch for a project that has later on become award winning, and the narrative would be incomplete without it.

Task Bar 8 allows for UK Government Information Exchanges. This task bar will not be necessary on all projects; it is included to ensure that UK Government requirements are met when appropriate. More information on how this data will be collected and used can be found at www.bimtaskgroup.org or www.thenbs.com

One of the fundamental benefits to the RIBA Plan of Work 2013 is the framework that it sets up for comprehensive and logical progression through the project. At the completion of Stage 2, the Information Exchange demonstrates to the client the work done and should be presented in a way that allows client to easily sign off the work to date and allow Stage 3 to commence.

CHAPTER 02

SUMMARY

- A comprehensive and collaborative design team response is developed to the Initial Project Brief from Stage 1, culminating in a Final Project Brief and a Concept Design.
- All Project Strategies are reviewed and developed for the stage completion.
- The Project Execution Plan is reviewed and updated accordingly.
- The Project Programme is reviewed and updated accordingly.
- The project Cost Information is reviewed and updated accordingly.
- Documentation for the Stage 2 Information Exchange is agreed, including the digital strategy.

SCENARIO SUMMARIES

WHAT HAS HAPPENED TO OUR PROJECTS BY THE END OF STAGE 2?

Small residential extension for a growing family

Although there has been some agreement that other design team members will be appointed during the design process, the architect has been working by themselves throughout Stage 2. They have produced a number of design iterations to illustrate some major and some more minor choices that the client will need to make. It became clear to the architect as some of the detail began being discussed that the adult couple of the family did not agree on everything. The architect has convinced them for future stages to make decisions as a 'single voice', and to speak through the architect to other consultants and, eventually, to the contractor.

In assembling the Stage 2 Information Exchange, the architect has prepared plan and basic main-elevation drawings to demonstrate size and general appearance. An outline cost has been produced using historic square-metre data for domestic projects of this size, and a series of suggested adjustments has been scheduled in order to assist in staying within the Project Budget figure. A Sustainability Strategy has been prepared, with comparative information on zero-carbon technologies. In addition, a simple 3D computer model has been built to demonstrate how principal views internally and externally will appear.

The client did not have any major concerns in relation to the Project Programme at Stage 0 and it had been developed to illustrate a more precise sequence of events and likely periods of time taken to reach significant milestones – and also to give a timetable covering when they will need to be ready with information themselves. For example, it records their last opportunity to amend the project brief, advise on kitchen-supplier preferences, or choice of light fittings, finishes and colours.

STAGE 2
CONCEPT DESIGN

Development of five new homes for a small residential developer

The developer has appointed the architect and structural engineer on direct appointments. They are very keen to reduce any risks in the ground, and have commissioned the engineer to undertake site-investigative surveys so that the foundation design can progress and robust cost information be developed for the substructure and ground works. The developer has their own quantity surveyor, who is going to develop and control cost information throughout the project.

Developing the Stage 1 Project Programme and procurement strategy has helped the client to decide to submit for planning permission at the end of Stage 2, and that a traditional form of contract is most likely to deliver the desired high-quality results. The architect has advised that submitting an application at this stage is likely to result in a significant number of planning conditions being imposed, but the client's interest is in securing a planning permission before committing to further expenditure. The client has decided not to undertake a pre-application process, relying on recent successful residential applications to indicate what will be acceptable to the planning authority.

The architect and client visited three similar-sized developments locally in order to inform a discussion on the local housing market and how the design needed to reflect the high aspirations of the target purchaser, which, in turn, would maximise the value of the development to the client.

Refurbishment of a teaching and support building for a university

The design team have been appointed up to the completion of Stage 3. The procurement strategy makes provision for that appointment to be extended, but intends that the design team will then be novated to the design and build contractor. The Design Responsibility Matrix (DRM) reflects this, leaving the post-novation DRM to be agreed later with the contractor. The university have introduced their health and safety advisor, who has established a risk register and circulated a schedule for their intended health and safety audits. The university also has, in addition to their original scope of services, asked that the design team work within BIM protocols for the rest of the project. This is to be a pilot project for the university, and they intend that a building information model should be the principal tender documentation tool at the end of Stage 3.

The architect has prepared a scheme that closely reflects the Initial Project Brief requirements, and has provided for slightly more teaching space than anticipated. The Cost Information produced at

Scenario C continues overleaf →

New central library for a small unitary authority

Stage 2 is a little higher than the Project Budget, but the client has agreed that this can be resolved during Stage 3 when a financial appraisal can be adjusted to take cognisance of the additional teaching space.

Each of the design team consultants has prepared separate reports on the existing building and the refurbishment scheme with recommendations for elements of work, which have been factored into the cost model. These reports – together with the architect's scheme drawings; the Project Programme; and a series of short Project Strategies, including a Sustainability Strategy – form the Stage 2 Information Exchange. A risk workshop is planned at the beginning of Stage 3 to help identify mitigation measures for the Developed Design stage.

A thorough and necessary review of the Stage 1 Information Exchange has been useful, together with a field-study trip to five similar-sized library projects funded in the same way and each with very high-quality design aspirations as part of a regeneration effort. The leader of the council is going to be the project champion.

The early site-investigation surveys have allowed the structural engineer and architect to work up a site-capacity model that demonstrates how a smaller than expected ground-floor footprint can deliver the appropriate project brief spatial requirements, and this in turn has delivered a more detailed cost plan than might normally be expected.

At Stage 1 the procurement strategy developed with the client determined that cost certainty together with a quality design project was paramount. To achieve this, they decided on a two-stage design and build Building Contract, with the contractor gaining preferred contractor status during Stage 2 and being appointed at the end of that stage. It is also anticipated that the client's architect will be retained by the unitary authority to monitor the contractor's design team and their Stage 4 design development. The procurement strategy also contains a BIM execution plan, setting out expectations of the design team and contractor, establishing project protocols and scheduling Information Exchanges for each Plan of Work stage.

The unitary authority had chosen this new library project to be an exemplar sustainable development in its ambition to be a leader in this area. The Sustainability Strategy at the end of Stage 2 responds to a very ambitious target of BREEAM 'Outstanding' (with a score of over 95%) included in the Sustainability Aspirations and includes a carbon-positive energy system utilising geothermal energy over 24 hours, photovoltaics and the highest standards of airtightness and build quality.

The Stage 3 element of the Project Programme reflects the significant amount of activity that will need to take place in order to produce robust tender returns, and the Information

STAGE 2
CONCEPT DESIGN

 New headquarters office for high-tech internet-based company

The digital entrepreneur and CEO of the company is to be the project champion. They have been instrumental in appointing a well-known architect and a design team that has established design credentials. The Stage 1 Initial Project Brief and accompanying Project Strategies have set a high bar for an award-winning sustainability-exemplary and user-friendly headquarters building.

Rather than setting a strict Project Budget from the outset, the client is concerned with value for money and has developed a project matrix that is intended to measure design quality, time and money factors against data sets for equivalent buildings. The design team are excited by the possibilities of their design work benefiting from this balanced view, and have scheduled the Stage 2 Information Exchange documentation to assist this measurement.

The Concept Design progresses well and the client has selected a preferred option from six prepared by the architect. The lead designer has progressed the strategic coordination issues with the rest of the design team and the project is going well. A management contractor has been appointed following a tender process and with a grid established and the building location on the site determined the preparation of the first packages for construction is underway. A number of façade options were prepared and the client has provided a clear direction on the way forward.

Exchange is supplemented with an expected information schedule for Stage 3.

The contractor is appointed on a pre-construction arrangement with sufficient time to contribute to buildability aspects and prepares the Construction Strategy that is aligned with the Concept Design.

CHAPTER 03

STAGE 3
DEVELOPED DESIGN

DESIGN
A PRACTICAL GUIDE TO RIBA PLAN OF WORK 2013
STAGES 2 AND 3

RIBA Plan of Work 2013

Stage 3

Developed Design

Task Bar	Tasks
Core Objectives	Prepare **Developed Design**, including coordinated and updated proposals for structural design, building services systems, outline specifications, **Cost Information** and **Project Strategies** in accordance with **Design Programme**.
Procurement Variable task bar	*The Procurement activities during this stage will depend on the procurement route determined during Stage 1.*
Programme Variable task bar	*The RIBA Plan of Work 2013 enables this stage to overlap with a number of other stages depending on the selected procurement route.*
(Town) Planning Variable task bar	*It is recommended that planning applications are submitted at the end of this stage.*
Suggested Key Support Tasks	Review and update **Sustainability, Maintenance and Operational** and **Handover Strategies** and **Risk Assessments**. Undertake third party consultations as required and conclude **Research and Development** aspects. Review and update **Project Execution Plan**, including **Change Control Procedures**. Review and update **Construction** and **Health and Safety Strategies**. *During this stage it is essential to review the **Project Strategies** previously generated.*
Sustainability Checkpoints	• *Has a full formal sustainability assessment been carried out?* • *Have an interim Building Regulations Part L assessment and a design stage carbon/energy declaration been undertaken?* • *Has the design been reviewed to identify opportunities to reduce resource use and waste and the results recorded in the Site Waste Management Plan?*
Information Exchanges (at stage completion)	**Developed Design**, including the coordinated architectural, structural and building services design and updated **Cost Information**.
UK Government Information Exchanges	Required.

CHAPTER 03

OVERVIEW

This chapter lays out the characteristics of the Stage 3 Developed Design process. This stage seeks to interrogate and develop the Concept Design through more detailed knowledge of the building materials and systems to be employed, and to understand how their cost and construction criteria might affect the programme. At the completion of the stage, the design team will present a range of updated and new detailed Project Information to enable the client to sign off the stage confident that the project parameters are deliverable and will be met.

APPROACHING DEVELOPED DESIGN

After the creativity and freedom often characteristic of the design process during Stage 2, the Developed Design stage can feel like a series of proofs for the propositions made during Concept Design. The design processes employed during the Developed Design stage require disciplined and coordinated scrutiny across the design team in order to ensure success against the parameters set for the project. This stage also involves considerable iteration in pursuit of a design that carries the concept forward but that also prepares the components of the project for the rigours of Stage 4 Technical Design in a way that avoids needing to amend the Stage 3 design to accommodate new factors.

It is within RIBA Plan of Work Stage 3 that the efforts of the design team really become fixed into place, and the various Project Strategies set up in earlier Plan of Work stages converge to make the coordinated design more rigorous and easier to achieve. We have seen in Chapter 2 that a planning application may have been programmed in at Stage 2 for a project- or client-specific reason. For many projects, however, the planning task sits more naturally within Stage 3. At this stage, the project will have reached a point in its development at which many more significant decisions will have been made and coordinated with the outputs of the whole design team. The client will also be able to really see the qualities and detail of the building that they have commissioned, and will have confidence in the design and cost information that is supporting the design. Planning applications made at the end of Stage 3 are significantly more robust than earlier-stage submissions, the designers having had the opportunity to produce the kind of detail that planners and the public are looking for from their built environment.

STAGE 3
DEVELOPED DESIGN

The planning process places the project into the public domain, and all the documentation that accompanies a planning application becomes publicly accessible material. Despite a pre-planning application process, which seeks to minimise areas of concern or difference between the project team and the planning authority's team and policy documentation, this is an uncertain time in the project's life. The planning system and its decisions are controlled initially by local, democratic processes, and, although a variety of preparations can be made prior to a determination of the application, the progress of the project beyond this point rests with the political leaders of the locality.

WHAT IS STAGE 3 DEVELOPED DESIGN?

The Developed Design forms Stage 3 of the RIBA Plan of Work 2013 and represents the culmination of the resolved and coordinated design. It is the fourth stage of the Plan of Work framework and, in this Stage Guide series, sits with Stage 2 Concept Design as one of the two stages at which significant iteration occurs in the creation of a building project following the Initial Project Brief but before the technical detail of Stage 4. The completion of Stage 3 is often regarded as the threshold between creating and delivering a project.

What is a Developed Design?

The ideas that are generated during the Concept Design stage show creative responses to the project brief: sometimes loose and exploratory, experimental without exact scale or the restriction of exact dimensions, but resulting in a cogent response to the client brief. The Developed Design stage focuses on rationalising and validating the detail of the decisions already made in order to allow the design team to be able to prove that the design works as a whole and checking further its affordability. This involves dimensional accuracy, checking elements of design against the appropriate regulations and ensuring that the actual construction requirements are not going to affect any assumptions made during previous stages. The client clearly has an interest in understanding the balances established by the design team in reaching this stage, and the coordinated effort necessary to ensure that the design team disciplines can each sign off against their own areas of expertise. It remains very important at Stage 3 that the design team continue to look ahead, so that they can clearly understand how the stages completed up to this one can contribute to the success of the whole project. The client should be aware of progress through this stage, but does not need to be directly involved unless some revision of the Concept Design conclusions becomes necessary.

COORDINATING DESIGN

It is no accident that when teaching building-design disciplines of all types, tutors will reinforce the need to account for the role and output of all the other design team members within this Developed Design stage of the process. Of particular importance to successful Project Outcomes is the coordination of structural and services information with each other, and with the prime internal building elements and external envelope. It is critical that each design team member is afforded sufficient time and opportunity to contribute to the delivery and management of what becomes the core stage of design coordination.

At Stage 3, the coordinated design ensures that there is dimensional fit and spatial compatibility between elements of the project designed by different design team members. Depending on the Technology Strategy for the project, this may be handled as a manual check or the *clash detection* module of a building information model.

Below are some areas of design coordination that will be established, and which should be anticipated during Stage 3.

- Confirm the structural-grid dimensions in plan at each level of the building. Do they offset at all, creating differing relationships with building elements on different levels?
- Establish the overall depth of the structural zone, and how often in each plan direction this depth is reached. What are the resultant restrictions on floor-to-ceiling heights?
- Consider what the relationship between structure and envelope is going to be, and what effect this might have on internal or external appearance and finishes. Does the structure 'read' on the outside, or will there be awkward corners and encasements in areas where hard finishes, such as tiling, is proposed?
- Confirmation of the main building services' vertical and horizontal distribution routes. Are they sensibly located for economic installation? Do they follow acceptable routes through the building? Are there any acoustic issues connected with service distribution that would affect areas of the building in use?

definition box continues overleaf ⤳

> **COORDINATING DESIGN** *continued*
>
> - Establish the depth of the building-services zone, and the parameters for changes in direction of services. Bends in services, and services changing direction to avoid structure, need the correct spatial fit; reducing the number of times that this happens can help reduce costs too.
> - Establish whether the structure can accommodate building services running through it, and at what centres. Is there an opportunity to create a single zone, within which a coordinated structure and services solution might work?
>
> Consider where building services will need to penetrate the internal walls and external envelope. The size and location of services terminating on external walls and roofs need coordinating with elevation materials and arrangement, in order to avoid undesirable effects on the building's appearance.

What is the delivery of Concept Design in detail?

Concept Design creates an intention. For example, in our project Scenario E, in order to create an 'industrial aesthetic', the architect wants the structural steelwork junctions and air-handling ductwork to be on show below the floor soffit. They know that to pull this off a high level of coordination is necessary, and that both structural and services engineers need to 'buy into' the concept as it involves them also thinking about what their elements of design will look like. In another example for the university building in our Scenario C, large sections of wall space need to be left free of services equipment or structure in order to act as display and projection space. The services design will need to respond to this requirement by avoiding switches, thermometers, alarm sensors, etc. on these walls, even though a standard design solution would inevitably seek to place some of these items on them.

The design element in question might be quite substantial – like a 'concept I' that has been conceived as part of a relevant street scene, responding to an Initial Project Brief that required a 'civic' presence. The choice of materials, the detailing for weathering, the marking of openings and expansion joints all contribute to the eventual impression of the façade. Developed Design proves that the delivery of these ideas in reality will contribute to the whole design, as had been envisaged during Stage 2.

STAGE 3
DEVELOPED DESIGN

During the Developed Design stage, other design team members become more actively engaged in responding to the spatial and environmental parameters of the building, and this is where the rigorously written and reviewed Project Strategies come into their own. A well-constructed strategy sets out the requirements of the brief, the condition under which those requirements might or must be met and any existing or presumed constraints that need to be taken into account. The Concept Design will have illustrated how these Project Strategies are most likely to be met in conjunction with each other, and when they subsequently come under review during the Developed Design stage it will be clear how the design team must work to deliver these coordinated elements of the building project. Clear Project Strategies will be able to relate these design intentions, whether it is the same designer working on the project or new members of the design team.

CLASH AVOIDANCE USING BIM

A key component of the coordinating design task is avoiding clashes between building elements which if remaining undetected can cost time and money on site modifying or changing aspects of the design. Typical clashes of building components that require detection at this stage might include

- Underground services and foundation arrangements
- Structural beams and horizontal services
- Structural bracing and external wall elements, including doors, windows and insulation
- Vertical service risers and structural and architectural components
- Bulkheads over stairs and other areas where head height is critical
- Roofing components, particularly awkward geometry or curved elements with structure and building services.

Building Information Modelling (BIM) and the emergence of 3D software tools has made clash detection much simpler and at Stage 3 allows modifications to take place without consequences on the building contract. The construction industry is aiming to achieve Level 2 BIM (Managed 3D model environment for separately produced discipline models) for publically procured works by 2016. Level 3 BIM envisages a single model worked on by the design team where clash avoidance happens during the normal course of the design developing.

DESIGN
A PRACTICAL GUIDE TO RIBA PLAN OF WORK 2013
STAGES 2 AND 3

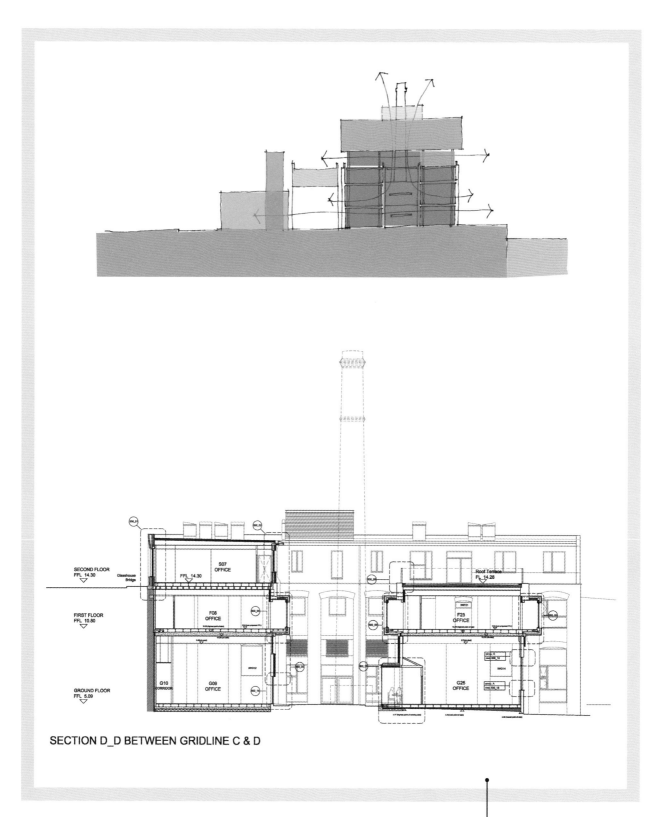

3.1
A sketch comparison of a Concept Design and Developed Design from the same project.

STAGE 3
DEVELOPED DESIGN

DESCRIBING THE PROJECT

All project team members will be familiar with describing the building project to a range of stakeholders. The language, expressions and detail used to do this change, of course, depending on who the audience is. These descriptors are important as part of the Information Exchanges that happen all through the project. Obviously, verbal exchanges are not often recorded, and design dialogue of this nature represents a high proportion of the development time spent between design team members exploring and modifying the building design.

So a Stage 2-type statement like 'the building envelope will be made of brick' is enough to evoke a sense of that design decision, but during the Developed Design stage the architect might add that 'the simple "punched" hole in the brickwork is lined with an aluminium flashing to four sides with the window frame set back from the face by 250 mm'. This better understanding of how the brick will form each window opening and how it is finished will add detail, complexity, time or cost to the project as a whole.

Also, it should be possible to be clear about the dimension, performance in use, availability and cost of the brick without writing a full specification for it. If as many elements of the building's construction as possible are identified with this level of information, the Developed Design will evoke a substance and quality in all respects that will survive the investigative rigours of future stages.

WHAT EFFECT WILL PROCUREMENT decisions have on Stage 3?

As discussed in Chapter 2, the procurement of design and construction services for building projects will vary depending on the choice of Building Contract and the client's priorities for the three principal components of any project – quality, time and cost.

The impact of procurement is tracked in the RIBA Plan of Work 2013 under Task Bar 2. This is a variable task bar that can have very different levels of activity in each stage, depending on the decisions made at the initial briefing stage. The amount of work required from the design team across all Plan of Work stages changes very little between procurement routes – the difference generally lying in who the Information Exchanges are prepared for, and the level of engagement of the design team with the client or end user in preparing the design through Stages 2 and 3.

> **PROCUREMENT CHOICES FOR COST**
>
> In simple terms, the later a project is tendered the more reliable the price will be, as it will be based on Developed Design and possibly Technical Design detail. Maintaining project team control over the design usually gives the highest design-quality results. This process tends to take longer to achieve, so some forms of contract seek to fix a price much earlier in the Project Programme. However, the earlier the works are tendered the less reliable the price is, as it is based on undeveloped design information and will contain client contingencies and provisional sums or the contractor's assessment of the risk remaining in the project.
>
> In all situations, the client's willingness to retain risk or requirement to transfer it to the contractor will dictate both the procurement choices and the extent of eventual control over cost decisions. Cost certainty does not mean cheap, and the loss of control over design is often very hard for clients to accept.

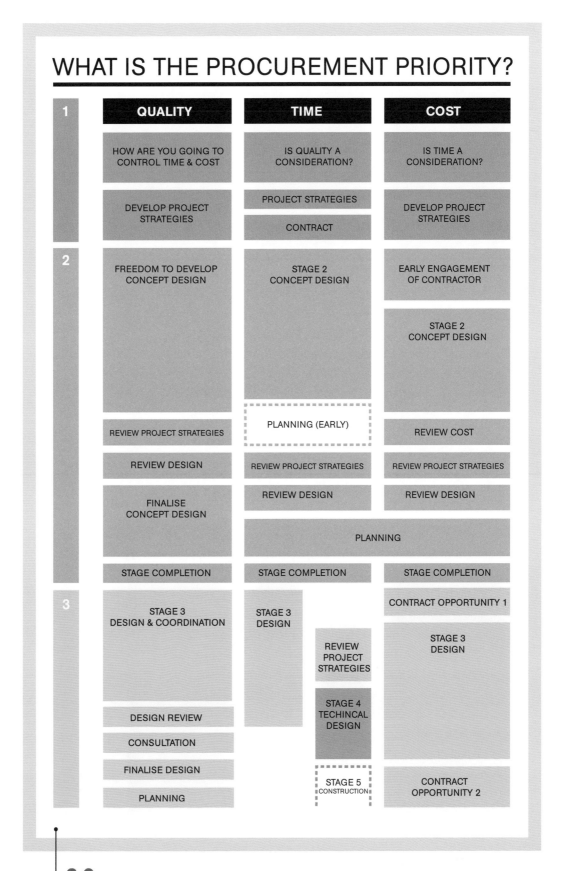

3.2
An extract from a procurement choice flow diagram relating task bar activity to Plan of Work stages under different procurement priorities.

The choices made for procurement of the building project will be fundamental to the nature of relationships in the project team. These choices do not affect the sequence of work described in the RIBA Plan of Work, or the type or level of completeness of the technical design information required. They can affect the information to be included in information exchanges, but do not diminish the need to put in place Project Strategies, information protocols and cost information at each stage.

THE IMPACT OF PROCUREMENT ROUTES ON STAGE 3

- Traditional – the Developed Design will assist in pre-tender cost estimates, and allows the client to remain in control of the design process. This level of control can change due to external factors, but generally means that the design team are able to focus on design quality and clarity of information. The Stage 3 Information Exchange clearly sets out how Stage 4 is going to deliver tender information in order to provide a credible construction cost.
- Design and build, single stage – most single-stage procurement is tendered at the end of Stage 3, when the design is fixed but the expertise of the contractor and specialist subcontractors can bring benefits to the construction and cost process. Information at this stage has to describe the full design intent even when it is not all declared in drawings and specifications.
- Design and build, two stage – the two-stage tender process is most useful when the contractor and specialist subcontractors are able to contribute their expertise in parallel with the design team during Stage 3. An agreed construction cost at the end of Stage 3, with this level of contractor input, before a building contract is signed contributes significantly to maintaining design quality in the finished building.
- Management contract – Stage 3 information will be developed concurrent with Stage 4 packages, in line with the expectations of the. For this type of procurement programme, the lead designer maintains the coordination of all elements of the project.

For more information on this subject, please refer to Construction: A Practical Guide to the RIBA Plan of Work 2013 Stages 4, 5 and 6 by Phil Holden – the third book in this series.

PROCUREMENT CHOICES FOR SPECIFICATION

Specification is the written part of the Stage 3 information output, which describes what materials, building products and systems the building will be made of. It can be produced with very different levels of detail depending on the procurement choice:

- Traditional – the design will be accompanied by an outline specification at Stage 3. This identifies in generic terms the materials and product types chosen for the project. This will be developed fully during Stage 4.
- Design and build, single stage – specification for single-stage procurement is included in the Employer's Requirements tender document. This will have been developed to include all elements of the project that are specific client requirements (eg a metal standing-seam roof from one of two manufacturers), leaving other areas as quite generic performance criteria (eg a high-pressure laminate WC cubicle system with easily available replacement parts), giving the contractor flexibility for particular products in their Contractor's Proposals document at tender return.
- Design and build, two stage – at the first stage of the two-stage tender process, the specification is likely to be quite generic, as it is in the traditional route. Working with the contractor through the second-stage process allows the specification to develop alongside the design and cost plan. At second-stage tender submission, these elements should be coordinated into a comprehensive Contractor's Proposals document.
- Management contract – the Stage 3 specification will be negated by the need to issue Stage 4 in packages. The lead designer and contract manager have to maintain an overview of each specification package, to ensure that interfaces are carefully considered.

How to deal with Stage 3 being the completion of the commission

There are certain circumstances in which Stage 3 Developed Design will be the completion of the design team's appointment. This may have been made clear from the beginning, because the appointment was to submit a design to gain planning permission for a site. It may be the result of a procurement route required for the project by the client or a funding agency, which means that design team services are competitively tendered ahead of Stage 4. It may alternatively result from the Building Contract, necessitating a contractor taking responsibility for the design from the beginning of Stage 4 and not requiring the continued services of the design team that had been responsible for the design up to that stage.

It is important to recognise that these circumstances occur considerably more frequently now than in the past. It is also important to recognise that if this situation might be the case for the project being worked on, then there is an additional responsibility on the lead designer and the design team to complete the Stage 3 Developed Design Information Exchange in a manner that can be easily understood and continued by the subsequent design team.

STAGE 3
DEVELOPED DESIGN

HOW DOES THE DESIGN PROGRAMME
in Stage 3 differ from that in Stage 2?

At Stage 2, the Design Programme organises elements of the emerging design in relationship to one another, resulting in a strategic Concept Design. During Stage 3, the Design Programme relates directly to coordination of the activity of the design team. Encouraging design declaration and dialogue to avoid conflicts in building elements results in a fully coordinated Developed Design at the Stage 3 Information Exchange.

The Design Programme will be the responsibility of the lead designer, and during Stage 3 it sets out appropriate time periods for key aspects of the Developed Design, design team and client meetings and design workshop events (discussed in Chapter 2). It can be used very effectively by the lead designer to encourage successful design team collaboration during Stage 3. It will culminate in an agreed and coordinated design that goes forward for approval to the client. The Design Programme will most likely at this stage include planning application and Building Control procedures, and may also include third party consultation or other stakeholder events that may impact on the design process.

The reason for the change of emphasis on collaboration in the Design Programme during Stage 3 is twofold:

- Information at Stage 3 is used to demonstrate that the Concept Design can be realised. Real components and construction products are brought together by different design team members to form a building that captures the essence of the Concept Design and can be costed. It is important to avoid change after planning permission has been granted and as the project enters into Stage 4.
- At Stage 3, time is the element of the project most likely to have defined parameters. At the outset of the project it is relatively easy for a client to say, irrespective of the project's scale, 'I'd like an x building project, built for £y by z date.' Unlike cost, which can fluctuate during the development stages, time is a linear factor in the project and a day lost is lost for good. This fact makes the management of the Design Programme during Stage 3 a critical contribution to the success of the Project Programme.

DESIGN
A PRACTICAL GUIDE TO RIBA PLAN OF WORK 2013
STAGES 2 AND 3

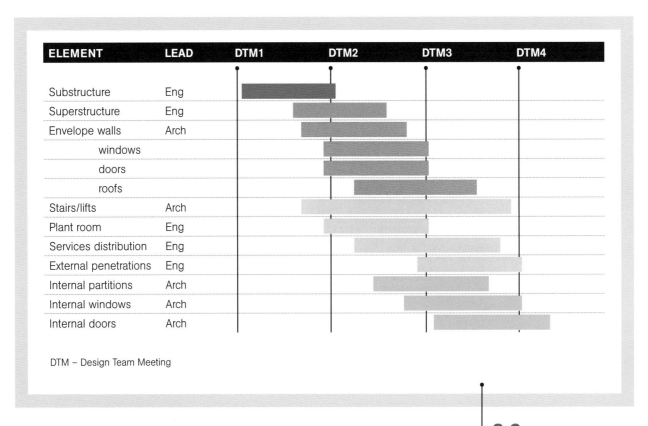

3.3
An extract from a Design Programme.

As the project enters Stage 3, the key project milestones that represent significant financial commitment from the client are coming into sharp focus. Typically, a planning submission fixes the design scope, and a start of construction on site is programmed in with a handover date for the completed project in view. By comparison with Stage 2, the Stage 3 Design Programme feels targeted and very time driven. Meeting deadlines to maintain programme, working collaboratively with other design team members and producing a coordinated design will result in a qualitative Information Exchange on time with good Cost Information and significantly less prospect of future design change.

STAGE 3
DEVELOPED DESIGN

WHAT DO YOU NEED FOR A PLANNING SUBMISSION?

(Validation checklists, information schedules, resource planning)

For the majority of projects, Stage 3 will include a planning application submission. This decision will have been taken during the Stage 1 briefing activity and be recorded on the Plan of Work Variable Task Bar 4, (Town) Planning. Depending on the nature and scale of the project, a significant period of formal pre-application engagement might be dedicated to meeting with and understanding the planning authority's opinion, planning policy and submission requirements. These activities will be recorded on the Design Programme, as their outcomes affect the iterative design process and, consequently, potentially affect the programme. Each planning application will be made up of validation requirements from a national and local list. It is good practice to formulate a validation checklist at the outset of Stage 3 in order to inform the scope of work, responsibilities and timescales for each piece of work, as they will inform other parts of the scope. Validation requirements are often set out in a checklist format, and usually available as a downloadable document from the local council's website.

These validation checklists are very likely to include a variety of specialist reports, which might form part of the planning-submission package depending on the scale and complexity of the project. If these required items are missing, the submission will not be registered as a 'live' application and, potentially, delay might be caused to the programme. It is quite usual for urban planning applications to be accompanied by the following reports:

- Desktop Site Investigation – the site conditions are recorded using historical maps to assess previous development; archive information on utilities, mineworkings or other licensed underground activity are logged in order to be able to consider site conditions for foundations and other civil-engineering structures.

- Geotechnical surveys, including contamination assessments – intrusive ground surveys using boreholes, trial pits and window sampling to confirm load-bearing capacities and contaminants.
- Ecology surveys and tree surveys – recording biodiversity and ecological value, and identifying species of flora and fauna.
- Heritage Impact Statement or Conservation Area Statement – considering impact of proposals on listed buildings or structures in Conservation Areas.
- Planning Statement – planning policy referred to as justification for the scheme being acceptable.
- Transport Assessment (for large projects), Transport Statement, Green Travel Plan outline.
- Noise or Vibration Survey – recording of noise or vibration factors that may affect proposed use.
- Statement of Community Involvement – it has become common practice to organise a consultation or engagement event with the general public before a planning submission. Comments and actions leading from those consultations are recorded in this statement.
- Design and Access Statement – an explanation of site context, design concept and design development, and proposals for accessibility across the project. Information produced during Stages 2 and 3 and have informed the design should be included in this statement.

Rural or suburban applications might require some of the above and the following reports:

- Sequential Test or Town Centre Assessment – to resist unnecessary out-of-town development and ensure sustainable settlements.
- Visual Impact Assessment – consideration of impact of development on established or relevant views.

It is worth noting that this list in effect becomes the majority of the contents of the Information Exchange at the completion of Plan of Work Stage 3 – and it may be worth considering the Design and Access Statement as the 'organising document' for the design information and these various reports, acting as one document doing two jobs.

The point to make here is that each of these reporting requirements will take time, and they must be planned appropriately within the Design Programme. Project Strategies are ideal for identifying what documents should be commissioned and when. Some of them will respond to project proposals, as they are reporting on the scale of impact of the proposed scheme (for example, the ecologist's assessment of wildlife habitat impacts). But some need to be carried out earlier, as they will impact on the design proposals (for example, the Site Investigation might identify an area of the site not able to be economically developed due to poor load-bearing capacity).

These reports should form part of the formal pre-application process, so that any impacts identified can be mitigated in the scheme design before proceeding to a planning-application submission. Careful and early consideration of these validation requirements will inform the Stage 3 Design Programme and help to ensure adherence to the target submission date for planning.

REVIEWING STAGE 2 PROJECT STRATEGIES

There is mention earlier in this chapter of the need to review some Project Strategies during Stage 3. All Project Strategies should be reviewed during this stage, even if some require only minor or no adjustment in order to remain relevant. The aim of the design team should be to complete the stage with a fully aligned set of Project Strategies, with changes from the previous stage highlighted.

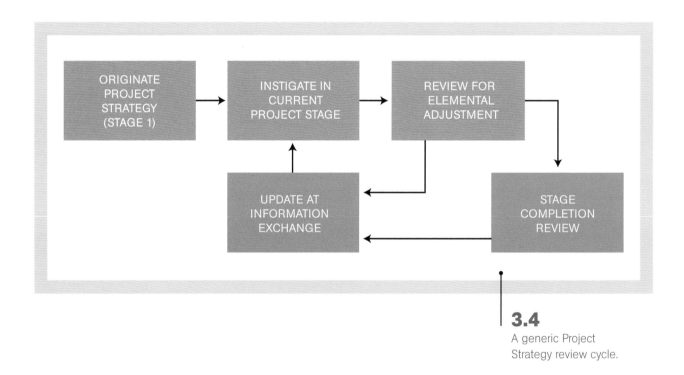

3.4
A generic Project Strategy review cycle.

During Stage 3, the level of detail that needs to be declared and understood is likely to warrant very significant updating of several Project Strategies that, in some instances, are linked. For instance:

- The Maintenance and Operational Strategy will be able to develop from a set of client requirements laid down at Plan of Work Stage 1 into a detailed set of provisions that the building design must allow for at this stage in order to avoid unwanted design change in later stages. For instance, the developed building design will make provision in its accommodation schedule for plant areas that during Stage 3 will be sized accurately, with layouts produced indicating the main incoming and outgoing services. Optimum locations for service risers, cleaner's equipment and general storage around the building will have been chosen, and each of these will inform the provisions of the Maintenance and Operational Strategy.
- The Health and Safety Strategy will be very significantly informed by the development of the plant room and roof-access protocols, and plant-replacement strategies, which all form part of the review of the Maintenance and Operational Strategy at this stage.
- These decisions together with design development in aspects of off-site assembly and manufacture will in turn impact on the Construction Strategy and the review of project Risk Assessments.

The review process illustrated in Figure 3.4 will be happening throughout the Stage 3 period, informed by the continued and qualitative development of the building project. The design team will need to be alert to the issues that impact on their sphere of operation, and perhaps influence the design decisions they make during this stage.

The Project Execution Plan will include a review regime for Project Strategies, setting out when the reviews should happen, who should attend review meetings or workshops, and the expected levels of output from these. For example, the Communications Strategy for the project will have established how digital communications and digital collaboration platforms will be utilised, and Stage 3 is a good time to review how that is working. The Project Execution Plan will be subject to review itself, and this process will help to record where changes are made across all Project Strategies and act as a directory of all activity during Stage 3.

How Project Strategies assist with statutory compliance

It should be considered useful to the design team to structure the relevant Project Strategies so that their contents are separated into, firstly, responses to statutory requirements and, secondly, those that respond specifically to project requirements or client initiatives. Typically, the extent of any Project Strategy that covers, for example, how any Health and Safety Regulations will be met may also touch on requirements that contribute to the safe operation and use of the completed building project. This distinction is important because it allows elements of proposed change or modification to the project proposals to be assessed against their impact on compliance on the one hand and project risk on the other. Clearly, the design team have a duty to avoid altogether any client exposure to non-compliance with regulations or legislation, and conscientious management and continued development of Risk Assessments during the design stages will guarantee this. Communicating the outcomes of Risk Assessments to the whole project team and maintaining a regularly reviewed project risk register is essential to good design development, and no less significant a task during Stage 3 than during the Stage 4 Technical Design development ahead.

OPPORTUNITIES FOR RESEARCH
in developing Project Strategies

In an ideal scenario, Project Strategies will develop in a linear pattern. Having set out at an early stage what the broad Project Objectives are in each specific strategy area, their development will follow the pattern of increasing knowledge around subject detail and emerging project criteria.

At Stage 2, the Project Strategies will have been reviewed in order to take account of new and emerging factors in the project – for example:

- The shape or massing of the building might affect the Sustainability Strategy.
- Site Information might influence the Construction Strategy or procurement strategy.
- The height of the building and the type of roof being proposed might impact on the Health and Safety Strategy.

As discussed above, these Project Strategies will be reviewed again, during Stage 3, in order to reflect the level of technical product and construction detail being investigated as the project become less generic. The key driver of this activity in Stage 3 is the need to increase the amount of fixed Project Information and to reduce to an absolute minimum the opportunity for principal project criteria to change during Stage 4 Technical Design.

During the individual and collective processes undertaken by the design team leading up to Stage 3, there invariably will have been opportunities for research projects to take place. In some cases, the development of new approaches, techniques or products that could emerge from the requirements of the project – and the desire of the project team to innovate or explore new possibilities – will have enhanced the project design outcomes at Stage 2. During Stage 3, the status of these research projects will be reviewed and possibly extended, provided that their Project Outcomes can be realised within the Stage 3 Design Programme.

This form of research and innovation in building projects is often associated with expensive schemes that can afford the prototyping and testing of significant new approaches to construction, or academic research projects that take time and have the objective of broad application in the construction industry. This scale of research often appears quite glamorous, but on occasion promotes very significant technological advances in building elements. These usually happen in conjunction with manufacturers and their product or process designers, who can see commercial benefit in the end product of the research. Much more common are circumstances within a building project that offer the opportunity for a piece of simple and inexpensive research. This might revolve around processes within a design practice rather than products that are specific to the project, but – if well set up, executed, recorded and disseminated – it will grow the knowledge base of the practice and the design team, and may have reputational advantage for the project and the design team involved.

IDENTIFYING AND UNDERTAKING SIMPLE PROJECT RESEARCH

1. A research project may emerge within a quite ordinary scenario – in response, for example, to a design requirement for a glass wall to perform in a certain way at dimensions that are unfamiliar to the design team, and when, while keen on the aesthetic value of the glass wall, the client has instructed that there must be no significant impact on the elemental cost plan for the building. On previous projects, the architect had only paid attention to the size of the window frames, ensuring that they fitted into whole-brick dimensions both vertically and horizontally, and so had never had reason to ask about glass manufacture. In the course of this project, however, the architect discovers, in conversation with the glazing supplier, the exact dimensions of the large glass sheets that are supplied prior to any cutting. The supplier invites the architect to visit their factory, where they are shown the process from supplied sheets to ordered product. It becomes clear that the number of cuts made to a supplied glass sheet, together with the waste generated

example box continues opposite ⌐

STAGE 3
DEVELOPED DESIGN

IDENTIFYING AND UNDERTAKING SIMPLE PROJECT RESEARCH *continued*

from those cuts, affects the cost per square metre of the supplied building element. This knowledge, together with a re-examination of design criteria, throws up the possibility of achieving the desired effect of a large glazed area while avoiding cuts to the glass, so that the relative coverage cost of the material does not affect the cost plan.

The architect writes a short research paper on this discovery, including an investigation into how this approach reduces material wastage from the process and how that might count towards an environmental rating for the building project. Clearly this knowledge can easily be shared, used on future projects and indicates the skill and dedication of the design team in wanting to achieve the briefed building project.

2. In another instance, the design team might seek to compare, on a much wider basis than normal, the relative merits of types of wall insulation, taking into account their whole-life and environmental costs rather than just their capital cost. In this building, the Project Programme is tight and of high priority – and one possible aspect of practice-based research could revolve around construction times for different forms of construction and the insulation they can accept, always taking into account the consequences for programming relative to cost. The outcomes for questions like these are not necessarily ground-breaking in an industry-wide context, but they can represent significant positive impacts within the project itself. An example of designing to cost

The seeming familiarity or simplicity of these project-based examples indicates that simple research of one sort or another happens more often than imagined. Following the framework of the RIBA Plan of Work 2013, and recording outcomes formally, helps to encourage research attitudes and allows the results to be shared easily. Such activity increases the knowledge base within the individual practice in the first instance, but then, more broadly, across the design or project team.

SETTING UP CHANGE CONTROL PROCEDURES
for this and future stages

On larger and, in particular, more complex building projects, Change Control Procedures may have been introduced before Stage 3. In all cases, it is a very useful project-control mechanism to introduce Change Control Procedures from the outset of Stage 3 particularly as the concept design and the aligned final project brief have been signed off by the client. Change control is a clear method for logging the following systematically:

- Changes to project criteria and detail.
- Who, or what circumstances, made the change necessary.
- The estimated cost of the change (design and construction costs).
- Any impacts on the Design and/or Project Programme.
- All the Project Information that needs revision in order to reflect the change.

A Change Control Procedure ensures that the client signs off the change before it is circulated to the design team to undertake the revisions necessary. It is good practice in a collaborative design team environment to have tested the proposed change with design team members, so that the impacts are widely understood and agreed.

STAGE 3
DEVELOPED DESIGN

PROJECT X_CHANGE CONTROL FORM

CCF reference:	xx/01
Date raised:	3/18/2014
Date required for sign-off:	3/21/2014

DESCRIPTION OF CHANGE
Client Team wants to alter ground floor reception areas and increase glazed frontage to street introducing automatic sliding doors

ORIGINATOR				
	Client	Y / N	Structural/Civil Engineers	Y / N
	Project Manager	Y / N	Health and Safety	Y / N
	Cost Manager	Y / N	Main Contractor	Y / N
	Architect	Y / N	Other (specify):	Y / N
	M&E Service Engineers	Y / N		

REASON FOR CHANGE		
Improved visibility and brand opportunity	Base Build	Y / N
	Building Control	Y / N
	Client Change	Y / N
	Design Omission/Coordination	Y / N
	Provisional Sum	Y / N
	Planning	Y / N
	Value Engineering	Y / N
	Other (specify):	Y / N

COST IMPACT				
	Professional Fees:	£500	Contingency	
	In-House Cost		Other (specify):	
	Main Contract	£7,500		
	TOTAL EXTRA:			£8,000

PROGRAMME IMPACT
3 weeks net impact on programme

DRAWINGS/DOCUMENTS ENCLOSED
Specification for new doors, revised Ground Floor and Elevational drawings attached.
Drawing Nos: 0000/200 01B, 0000/210 01A

IMPLICATIONS/CLARIFICATIONS/ COMMENTS		
	Time:	3 weeks
	Capital cost:	£8,000
	Business impact:	High

APPROVAL	Approved signature	Date:	DD/MM/YYYY	**R A G** circle
RED	Rejected		Instruction reference:	
AMBER	Approved with comments (attach comments)		Status:	
GREEN	Approved			

3.5
An example of change-control data template.

This mechanism of recording changes to the design, particularly during Stage 3, allows progressive design changes and cost increases and decreases to be tracked and monitored. During project meetings, the record of change control creates a useful history of change decisions, the reasons for the changes and a log of the authorisation for change. allowing informed discussion on the subject if required. This management element of the project is most likely to be encapsulated within the Project Execution Plan at Stage 1, along with a record of who takes responsibility for keeping the change control records up to date.

Developing and executing third party consultations

No building project exists in isolation, and in every case a series of third party consultations needs to take place in order to inform the project scope, cost and programme. Stage 3 is an ideal period in which to undertake these consultations, as the design team are able to say generally what the project proposal is and to absorb the information from the consultation into the Developed Design workflow. Third party consultations are noted under Task Bar 5 Suggested Key Support Tasks.

Here are some third parties commonly consulted during Stage 3, and what to expect from them:

- Amenity and civic societies – special-interest groups that may be active in the project location. For example, The Georgian Society, The Victorian Society or The Twentieth Century Society. By its very nature, the commentary received from amenity or civic societies will have a narrow focus and rarely seek to see scheme proposals in the round.
- Utilities providers – water, gas, electricity and telecoms services will all be delivered to the new project by different organisations. Each utility, in turn, may need to consult separate infrastructure, distribution and connections companies. Long lead-in times for new services often impact on the construction programme, and early knowledge of this can help establish the programme's critical path.
- General public – for around 30 years now, it has been considered good practice by many clients – particularly public-sector and quasi-public-sector bodies, like Housing Associations – to hold a form of consultation on proposed developments before they are submitted for planning permission. Listening to local people will inform design proposals, and

may mitigate formal objections during the planning application period itself.
- Design Review – CABE (the Commission for Architecture and the Built Environment) introduced a UK-wide network of Design Review Panels that provide an independent voice for design quality prior to planning. Design review commentary is usually extremely helpful at Stage 3 in identifying both strong and weak points in the design.
- Building for Life 12 – applicable to residential schemes, Building for Life produces a score out of 12 for the quality of place that a scheme promotes.

These consultations make take the form of relatively short presentations, longer exhibitions or workshops around specific interest groups. Such events can appear time-consuming when developing the Design Programme, but they do contribute considerably to the completeness of the Developed Design, informing the review of Project Strategies for subsequent stages, and therefore often save time resolving incomplete design thinking later on when there is less design flexibility. Avoiding pressure to resolve design matters in later stages also benefits the quality of the design, as decisions are made with appropriate information and time to consider options.

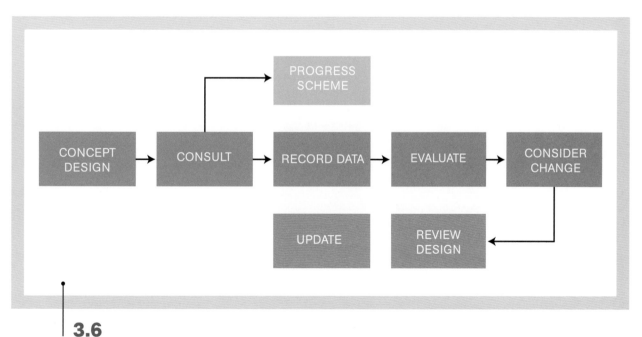

3.6
A process map for third party consultations.

EXAMPLES OF THIRD PARTY CONSULTATIONS

The following bullet points provide a brief illustration of some of the types of consultation events that even a modest building project might benefit from. Clearly, some of these events can relate directly to the continual review of specific Project Strategies and inform the stage-completion Information Exchanges of these. Examples of consultee organisations are given in each case.

- A user group representative of the people who will use the building and who have specific requirements that need to be catered for. This group might meet three or four times during the design process in order to share needs and provide comments on designs as they evolve – eg a tenants' association development subcommittee.
- A special workshop event with the whole project team represented in order to review accessibility issues – not only from a statutory-compliance point of view, but also with regard to whether any of these issues form part of the 'identity' of the building or a particular service that the client wishes to promote from the new premises. Some projects will have an accessibility consultant, who will lead on this aspect of the Developed Design stage (accessibility strategy) – eg the Royal National Institute for the Blind (RNIB).
- A design workshop with relevant design team members and client end user or facilities management representatives, if appointed, on maintenance and plant-replacement strategies; maintenance regimes, as far as they are understood; and contract cycles (Maintenance and Operational Strategy) – eg the appointed building caretaker or facilities manager.
- A design scenario workshop with relevant design team members and client end user or facilities management representatives, if appointed, to review how all daily operations, including reception function, will work; communications equipment and data installations; 'first

example box continues opposite ↗

EXAMPLES OF THIRD PARTY CONSULTATIONS *continued*

in/last out' security sequence and protocols for alarms, including personal-safety issues that might arise from after-hours working; CCTV and other security installations, including car-parking arrangements and the protocols and requirements of fire drills and evacuations; and protocols for alarm performance in relation to external and emergency services (Operational Strategy). There should be an opportunity for the comments and requirements recorded by the first of each of these design workshops to be illustrated within the design context and relayed back to the group for further comment – ideally, if time permits, within the Stage 3 Design Programme – eg the security contractor; the Secured by Design Architectural Liaison Officer (ALO), a member of the local police force who advises on security in designs.

- A public consultation can take the form of an exhibition or a more interactive event. The aims of the consultation should be discussed and agreed with the project team before the event. Comments and observations should be encouraged via an easy-to-use method, but it should be clear to members of the public attending how the client and design team intend to respond to their comments. How is feedback going to be delivered? Is there a publicity opportunity for the future project, of which this public consultation can form a part? What should the interaction with local ward councillors, planning committee members and officers, and other statutory consultees be at this stage? The planning application and the procedures that lead to a determination of that application are embedded in local democratic processes, and there is a need to understand – particularly in relation to controversial schemes or land-use proposals – how best to frame the supporting arguments for the project against the potential for, and content of, objections throughout the application period. Public consultations held prior to a planning submission can often act as a test ground for these opposing arguments – eg the developer, local councillor(s), residents' associations.

In the context of the RIBA Plan of Work 2013, it could be relatively easy to underestimate the powerful outcomes from a properly planned and executed consultation programme. Timely consultation of interested third parties creates a constructive working relationship, and informs the design process before the stage at which consultee input might create abortive work. On the other hand, a thoroughly worked-through Design Programme naturally seeks comment and critique on evolving design issues, and these opinions inform both the design process and the criteria that will eventually govern the building project becoming a success or not.

Consultation related to planning

Earlier in this chapter, reference was made to public consultation or engagement events held prior to a planning submission. It is also worth noting briefly the consultations that take place during a planning application period. Town and country planning legislation has embodied within it a statutory public consultation period of 21 days. This is notified on site, and gives any objector or supporter of the scheme an opportunity to make written representation to the planning authority – usually the local council but some areas are covered by a special authority, for example one of the UK's National Parks.

The same period is used for what are known as 'statutory consultations' with bodies such as the Environment Agency, Historic Scotland, Cadw, Northern Ireland Environment Agency (NIEA) or English Heritage, Natural Resources in Wales, Scottish Natural Heritage, NIEA or English Nature, who have the opportunity to say whether the proposed development scheme impacts on their jurisdiction or apparatus. Internal council consultees are also formally approached for comments; these will include departments such as Highways and Transportation, Environmental Health and Landscape, but may also involve other specialisms like conservation or the tree officer. This is a formal process as part of planning legislation, and is not discursive or exploratory in nature. If pre-application advice has been taken, it is likely that all the statutory and internal consultees will have already commented; however, once the planning application is registered there remains the opportunity for all comments received to influence the design before a permission is granted.

WHAT FORMAT DOES COST INFORMATION take at Stage 3?

A design team that understands the cost of construction and maintains an interest in developing Cost Information is well placed to complete Stage 3 with a design that remains within cost parameters, and with suitable design development allowances for Plan of Work Stage 4 Technical Design. Cost Information during this stage will inevitably rely on historical elemental cost data for construction types. This will produce a 'Construction Cost' estimate, and to complete the cost plan the cost consultant will use industry-standard indexed predictions on cost inflation for materials and labour, together with project-specific contingency sums against known project expenditure (for example, furniture and fittings) and some general contingency sums for possible future discoveries.

Cost Information at Stage 3 is usually presented as showing the main elements of the building (eg substructure, superstructure, building services). These elements will have varying levels of detail behind them, depending on the nature of the project and the extent of research undertaken into the cost of building components and products as described earlier. The level of cost detail gathered from design workshops, product manufacturers and other cost data will be variable, and the design team must work together to provide the best possible cost-planning information. Where detail of certain elements is less robust, the lead designer should work closely with the cost consultant to make certain that the contained in estimate allowances are realistic and appropriate.

AREA RATE (EXAMPLES) £/sq m

Building Type (including external works)
eg Residential £1,150/sq m; Art Gallery £2,550/sq m; Hospital £3,210/sq m

Building Only (excluding external works)
eg Commercial Office £1,350/sq m; Retail Shell £850/sq m

Project Costs (incl fees, project contingencies, fit out, etc.)
eg Commercial Office £1,850/sq m; Retail Shell £1,050/sq m

VAT
VAT can be charged at full rate, reduced rate or exempt rate. This varies according to project type.

Exclusions
These are usually items that will cost something but are indeterminate at this early stage and would misrepresent the project viability if they were included.

ELEMENTAL COST PLAN (ILLUSTRATION)

No.	Element of construction	Type	No.	Rate (£)	Elemental total
1	Substructure	linear metre (lm)	150	£25	£3,750
2	Superstructure (frame)	weight (kg)	3,500	£2	£7,000
3	Superstructure (walls)	square metre (sqm)	608	£300	£182,400
4	Building Services	itemised cost plan (lump sum)	1	£120,000	£120,000
5	Doors	number (no.)	3	£1,200	£3,600
6	Windows	number (no.)	8	£850	£6,800
7	Finishes	sq m	3,420	£35	£119,700
8	Contingency	number (no.)	1	£12,000	£12,000
	SUBTOTAL				**£455,250**
9	Preliminaries	% of Construction Cost	12	£4,553	£54,630
10	Inflation	% uplift	2.16	£5,099	£11,013
	TOTALS				**£520,893**

NB: All numbers are for illustration only and do not represent any real scenario

3.7 Types of cost data set up in a tabular form for cost planning.

Designing to cost at Stage 3

Cost is always going to figure as a high priority within the project criteria set out in the Initial Project Brief. The relationship of cost to the remaining two prime components of a project, time and design quality, will dictate some of the sequencing of tasks during Stages 2 and 3. Designing to cost has been discussed in the previous chapter, and in Stage 3 the relationship between design and cost is critical in delivering a robust Information Exchange that the client can have confidence will deliver their project to cost with the design quality illustrated at this stage.

THE IMPORTANCE OF SUSTAINABILITY
running throughout the project

Task Bar 6 on the RIBA Plan of Work 2013 makes provision for a sustainability checkpoint at each stage. Every building project will have a set of sustainability matrices mapped out at the Initial Project Brief stage, which will be measured during progress. Some of these may be funding requirements, client targets or external stakeholder provisions that need to be met. It is important to distinguish between those external goals and aspirations set out to allow the project to be realised from those that may be best-practice goals for the construction industry, personal or team targets or innovation targets, which do not have a bearing on whether the project proceeds. These two things though do come together to form the Sustainability Strategy.

It has been clear for a number of years now that new building projects have a substantial contribution to make to UK and global carbon-reduction targets, along with other environmental targets on resource usage. There are many ways of a project achieving a meaningful contribution to a sustainable future, and at the Stage 3 Information Exchange the methods committed to and means of achieving these targets need to be embedded in the design and cost information.

Stage 3 Sustainability Checkpoints from the Plan of Work ask:

- Has a formal sustainability assessment been carried out? If it has at an earlier stage, has it been reviewed during Stage 3?
- Have an interim Building Regulations Part L (conservation of fuel and power) assessment and a design stage carbon/energy declaration been undertaken?
- Has the design been reviewed in order to identify opportunities to reduce resource use and waste, and the results been recorded?

ESTABLISHING A SIGN-OFF PROTOCOL
for Stage 3 completion

Under an earlier section of this chapter, the possibility of Stage 3 being the end of a commission was discussed. It is important to also recognise the prospect of a significant time lapse occurring between the completion of Stage 3 Developed Design and the beginning of Stage 4 Technical Design. The reasons for this sort of delay may be attributable to general economic conditions, project funding procedures, occupancy of the existing site or building, or the planning process itself. The particular reason is far less important to the smooth running of the project than the ability of the design team to present all the design information gathered and produced during Stage 3 in a style that is both easily understandable and accessible in the future. This is the role of the Information Exchange, Task Bar 7 of the Plan of Work. The completed Developed Design, collated with all relevant information for the stage, should be organised so that it can be picked up, understood and immediately built upon at the commencement of Stage 4 Technical Design.

One of the reasons that this Information Exchange needs to be comprehensive is that individuals within the design team inevitably become embedded within the design development process, and carry a lot of quick-access knowledge about the project while working on it. Over the course of the time delay between stages, project staff may have changed projects within a practice, changed employment or otherwise be unavailable to use their retained knowledge to assist at the start-up of the subsequent stage. If they are still employed at the same practice, it may, of course, be possible to elicit some knowledge from an internal, discreet 'handover' meeting. However, this opportunity should not be relied upon, and in any case the written documentation made at the time of completion is going to be more reliable than the recall of individuals.

STAGE 3
DEVELOPED DESIGN

A good way of achieving this aim is to ensure that the stage completion information fits the following criteria:

- A narrative that illustrates the original conceptual intention, in order to avoid misinterpretation or loss of the Concept Design.
- Clearly drawn or modelled information with key dimensional data displayed. Each design team member will have the following information as a minimum:
  - ~ Architect – site layout; dimensioned plans, sections and elevations; some relevant larger-scale details to illustrate design intent; specification notes for all works.
  - ~ Structural engineer – structural grid and structural member sizes, foundation type and layout, underground drainage layout, specification notes for all works.
  - ~ Building services engineer – plant room layouts; distribution routes, sizes and equipment; heat and lighting layouts; 'Part L' energy-use model and calculations; specification notes for all works.
  - ~ Landscape architect – site layout, planting schedule, specification notes for all works.
- A BIM project should include a coordinated model with all the above integrated into a single model.
- Updated Project Execution Plan and Project Strategies.
- Cost Information related to the outline specifications above.
- A Project Programme demonstrating how time has been allocated to future Plan of Work stages.
- If necessary due to a break in programme, an explanatory narrative illustrating how final design decisions were arrived at, in order to avoid loss of project knowledge from any personnel change.
- An easily referred-to log of final decisions in the Project Strategies.

The Information Exchanges at stage completions require the discipline of design team members in order to provide good-quality Project Information in a timely and presentable manner. At Stage 3, this involves developing the project with well-considered and reviewed strategies; a fully coordinated design, with researched and confirmed Cost Information; and having a mechanism, as part of the Project Execution Plan, to allow a check against the Initial Project Brief. This should result in a successful project delivering excellent design standards.

As at Stage 2, the project team should consider the requirements of a UK Government Information Exchange if it is relevant for the project.

WHAT SHOULD BE IN THE INFORMATION EXCHANGE
at the end of Stage 3?

As just described, when the tasks in a stage are completed the design team prepares an Information Exchange for handing to the next stage. At Stage 3, this is a particularly important point in the project when the overall design 'feels' complete. The client signs off the design so that a planning application or tender process can commence. The Information Exchange at the end of Stage 3 Developed Design should include the following:

- Design information from all design team members. As noted earlier, this might include appropriately scaled and dimensioned plan, section and elevation drawings; photomontage illustrations; 3D electronic and physical models; a document detailing the site analysis, context and design constraints; and an appropriate level of specification detail relevant to the procurement route. The level of detail in a BIM environment should be sufficient to allow full coordination between architecture, structure and building services:
    - ~ establish main site and finished floor levels in relation to existing site levels
    - ~ detail main structural grid and principal plan and section structural dimensions
    - ~ detail main plant and equipment room requirements
    - ~ detail main service zones and distribution routes and how interface with structure is working
    - ~ detail building services external envelope penetrations
    - ~ detail main external wall and roof build ups
    - ~ detail all internal wall, floor and ceiling types and finishes
    - ~ detail door and window openings
    - ~ detail main external works elements.

    The coordinated design must be reflected in the Cost Information.

- The procurement strategy. The procurement route will be determined and embedded in programme information, and reflected in the Project Execution Plan. The procurement strategy will detail the information relevant to the chosen route, and this will form part of the Stage 3 Information Exchange.
- A Project Programme, showing the time period allowed for the remaining stages of the project.
- A Design Programme coordinated with the Project Programme, showing the proposed sequence and level of detail required during Stage 4 Technical Design.
- The Project Budget. This should include a construction cost plan and be clear about what is included and what is not. The cost and design must be aligned.
- Planning information: Details of the planning submission and a status report. Planning permission may be a prerequisite to starting Stage 4.
- Health and safety implications. These will have been considered, risks mitigated where possible by the design team, and a residual risk register prepared.
- All Project Strategy documents. These key strategies will cover Sustainability, Maintenance and Operation, Handover, Construction and Health and Safety.
- The Project Execution Plan will have been updated, and Change Control Procedures will be in use.

If during Stage 3, or earlier in some projects, it becomes a requirement to complete the building as fast as possible, the design team must immediately consider what the impacts on Information Exchanges are likely to be. As in our Scenario E, it is possible to conceive a Project Programme that overlaps the end of one stage with the start of the next. Additional diligence is required in progressing from Stage 3 to Stage 4 without a fully resolved design, as this increases the risk of coordination issues arising before or during the construction stage. Using the RIBA Plan of Work 2013 as a tool for mapping the sequence of tasks, even if overlapping, will promote good project management in this pressured situation.

EMERGING DIGITAL INFORMATION TYPES

A number of digital technologies are becoming influential in architectural practice as extended tools to explore concept designs and diversify the range of Information Exchanges. Many will be used during Stage 2 but arguably they will be most adeptly utilised during Stage 3. Some of these technologies might include

- Enhanced virtual realities and immersive environments where physical and sensory experience of the project design becomes a possibility
- 3D printing from simple 3D or BIM models – exploring building components, parts of or whole buildings. Relatively quick and easily set into context 3D printing
- Parametric design allowing software to create architectural form with given data sets for the programme or setting of the project
- Architects are becoming coders and scripting elements of the digital design toolbox to create unique approaches and identifiers within their designs
- Improved visualisation techniques and software design for photo-real CGI

Digital technologies are forecast to continue changing the design team toolbox at a very rapid rate, so while the processes behind the design will conform to the Plan of Work framework the outputs will continue to help the design team provide more accurate information.

CHAPTER 03

SUMMARY

- At Stage 3, the design becomes more detailed, accurate and coordinated with the outputs from the design team. The design will be clear in its scope and backed up by information that demonstrates that it is achievable within the project parameters set.
- The lead designer develops and manages the Stage 3 Design Programme, and reviews what is required during Stage 4.
- The design team reviews and updates all Project Strategies as necessary.
- The Project Programme and Cost Plan are reviewed and updated.
- The Project Execution Plan is reviewed and updated, adding Change Control Procedures if they have not been included at an earlier stage.
- A planning application is prepared and submitted, unless it has been previously agreed that it should be submitted at Stage 2. The design team should be aware that this action puts the project into the public realm for possibly the first time.
- A comprehensive and clear Information Exchange is prepared for the completion of Stage 3. A delay between stages or a change to the design team personnel is most likely at this point, and it is essential to the integrity of the project that the Developed Design translates successfully to Stage 4.

SCENARIO SUMMARIES

WHAT HAS HAPPENED TO OUR PROJECTS BY THE END OF STAGE 3?

Small residential extension for a growing family

The architect and structural engineer have completed a coordinated design for the residential extension. Taking the architect's advice, the client agreed to attend a series of pre-application meetings with the planning authority, which has allowed a planning application to be prepared taking on board a number of statutory-consultee comments. The planning application submission to the local authority forms part of the Stage 3 Information Exchange, and will be submitted via the online Planning Portal (www.planningportal.gov.uk).

A simple 3D block model has been developed in sufficient detail to be able to generate photomontage street views and garden views, in order to address client concerns about impact on neighbours and to accompany the planning application. The client has no interest in how the information for the project is produced in Stage 4 and recognising the limited coordination benefits for this scale of project in using a building information model the design team have decided to produce 2D information due to the tight timescales.

The architect has spent a lot of time during the Stage 3 Developed Design process discussing the relative merits of a number of external wall, window and roof materials, and a full spectrum of internal finishes too. Because the external materials are important for planning permission, the client has considered longevity, repair and replacement alongside appearance, and a detailed log of these choices has led the architect to instigate a research project on regional brick patterning, coursing and detailing. This has helped to assure the planning authority that the quality of detailing in the project will be very high.

The architect has advised the client that with a traditional procurement route the tender information needs to be fully detailed to avoid cost uncertainty or additional cost by late introduction of information or changes of design. In addition, the architect has expressed a concern about the client's Stage 1 decision not to appoint a building services engineer on the design team when a significant amount of services information

Development of five new homes for a small residential developer

was generated by the Final Project Brief and the client's Sustainability Aspirations, which will be requiring detailed Stage 4 information in order to assemble a robust tender package and pre-tender cost estimate. Following discussion, it is agreed that a services engineer can be appointed to close out Stage 3 and produce tender information at Stage 4.

The Project Programme has been revised at the end of Stage 3 to allow for minor delays during the pre-application process. Now that the Detailed Design is nearing conclusion the length of time needed to produce a comprehensive tender package is clearer, and the client is keen that Building Control full-plans approval is received before going out to tender. Although this produces a construction start four months behind that originally hoped for, the client's main driver is quality of the build and, as they intend to stay in the house during construction, no particular dates are placing pressure on the programme.

Planning permission was achieved at Stage 2 and, as the architect had advised, there were a greater number of planning conditions than normally expected. While progressing the Stage 3 Developed Design, the architect has had to resubmit elements of the design for minor amendments, and this has been able to 'capture' some of the conditioned information at the same time.

The developer has continued to drive down the budget for the scheme through amendments to the specification, and, while preferring a traditional contract, has indicated that they want to be on site at a point in the Project Programme much earlier than the design team can complete a robust tender package for this sort of contract. As the scheme is residential, the developer is prepared to work from their own historical data to create cost information that reflects the essential internal detail (joinery details, kitchens and bathroom fittings, etc.) and to include provisional sums against these items, allowing the design team to focus on the envelope design.

The lead designer, the architect in this project, has expressed reservations about this curtailed programme, but understands that this pressure comes from the way the project is financed and the desire for early sales revenue from the scheme.

The design team have prepared outline specifications and general-arrangement technical design in order to allow a Building Control submission. The client is only interested in achieving minimum regulations, but a Sustainability Strategy has been prepared to allow an initial 'Part L' model to be produced and to demonstrate how a 'Fabric First' approach to the design and careful orientation on the site contribute to the benefit of the occupants. The design team have convinced the client that lower energy bills and the beneficial use of solar heat gain can be used as sales incentives rather than being seen as unnecessary costs.

Refurbishment of a teaching and support building for a university

The client's financial appraisal of the Stage 2 Concept Design supported the inclusion of the additional teaching space created by efficient space planning during the Concept Design stage. Despite the scheme costing a little more than envisaged the revenue return exceeds the additional borrowing costs. The Final Project Brief has been adjusted accordingly.

As highlighted at Stage 2, the design team's appointment finishes at the completion of Stage 3. The client's intention had been not to commit beyond known funding support. Although this has now been received for the completion of the project, the project lead is still developing the procurement strategy with the client and no decision has been made in respect of whether the design team will be novated to the design and build contractor or not.

The design team have prepared the Stage 3 Information Exchange as for a single-stage design and build tender package, and this includes aspects where specialist subcontractors will have design responsibility, relevant Project Strategies, a risk register and a schedule of tasks, in the form of a Design Programme, to be completed during Stage 4 Technical Design by the contractor's design team. This last-named document was produced due to the uncertainty around the contractor's eventual design team. The Design Responsibility Matrix has been updated to include all the contractor's design responsibilities and to clarify how specialist subcontractors will contribute to the Stage 4 Technical Design process.

Also included with tender documents is a schedule of investigative surveys to open up the existing fabric, so that the contractor and their design team will have this information to establish a coordinated design. These surveys are to be costed as part of the contract, and provisional sums used against items of work that will be confirmed once surveys are complete. The items are recorded on the risk register, and the design team have highlighted to the client that there is a significant cost risk attached to this methodology.

Because of the considerable uncertainty at Stage 3 completion about exactly which way the design will be continued, the rather brief Communication Strategy from Stage 1 has been significantly enhanced in order to ensure that lines of communication between the eventual parties to the contract are well understood and easy to use.

New central library for a small unitary authority

During Stage 3, the client's design team have worked with the contractor's design team in a series of design, risk and operations workshops. Adherence to the initial Project Objectives was key to these workshops, and participants thought that they were a great success in delivering a comprehensive Stage 3 Information Exchange. The contractor has made significant contributions to the buildability of the scheme as part of the Construction Strategy, suggesting prefabricated elements of envelope construction, and, as a result, is offering an appreciable reduction in the Stage 5 Construction period of the Project Programme.

The two-stage design and build contract is in the last round of negotiations on lump-sum price, and the client's and contractor's design teams have agreed the Stage 4 Technical Design Programme. There are no changes to the Design Responsibility Matrix for this stage. There is a small amount of concern that the levels of risk contingency allowed in the lump sum are too high given the amount of early site investigation work that already exists. Negotiations between the client, cost consultant and contractor resolve this matter.

The BIM execution plan for Stage 3 is completed by the BIM manager (the client's architect on this project), and the protocols are accepted by the contractor's design team. The contractor during Stage 3 has developed their own BIM capability, and is planning on using the project as a pilot for how BIM can streamline their programme reporting – particularly around 4D (time) information and linking the model to the Construction Programme. Some additional fees have been agreed with their design team to help facilitate this task.

The Stage 3 Information Exchange includes a comprehensive Developed Design report that, subject to agreement on the contract sum, has been signed off by the authority. While an outline specification would normally be agreed by the client's design team at this stage, the fact that there has been a collaborative working relationship means that the contractor has confirmed 80% of the proposed materials and finishes – and this has assisted in cost planning.

The agreed Project Programme requires a Building Control application to be made as soon as Stage 4 begins, in order to allow commentary and changes to be absorbed within the Stage 4 Technical Design Programme. As the contract will have been let by then, the contractor's internal Change Control Procedure has been established to ensure documentation and correct approvals for any proposed changes.

New headquarters office for high-tech internet-based company

The management contractor has prepared a Construction Programme and a procurement strategy. The immediate impact on the Stage 3 Design Programme was to identify and begin assembling work packages to be tendered as soon as they are ready. The principal reason for choosing this form of contract was to allow work to start on site at the earliest opportunity, as the client has witnessed unprecedented growth during the previous six months and needs to be able to occupy their new headquarters as soon as possible. Out of approximately 15 work packages, the first two – substructure and concrete frame – are already out to tender, and a start on site is expected before the conclusion of Stage 4 information in other packages.

The client signed off the design at the end of Stage 3, and it has been submitted for planning permission. The client is aware of the risks involved in allowing Stage 4 design work to commence before approval is granted, but feels confident that consent will be granted and is willing to accept the risk.

The management contractor has produced a Stage 3 cost estimate for the project that is within +3% of the Project Budget. Some market testing has taken place in order to inform this cost, and a target price has been incentivised by the client. The Design Responsibility Matrix has been updated to reflect the complexity involved in multiple work packages, in particular highlighting where responsibilities lie at various package interfaces in order to avoid conflict later.

The Handover Strategy has been developed in detail during Stage 3, as the client wishes to undertake their own IT fit-out with various specialist equipment installations before Practical Completion. Partial possession is also being considered, to allow early occupation of up to four months before Practical Completion of almost a third of the building. The company is growing so quickly that the client has also broached the subject of adding a 30% floor-area expansion of the building to the contract if it is at all possible without causing delay to the main building. The CM and design team have quickly assessed this new prospect, and it has been decided to seek planning permission before the end of Stage 4 of the existing project – adding the additional work in work packages by way of a variation, and attempting to bring the whole project to Practical Completion at the same time. The Design Programme, Project Programme and Handover Strategy have all been updated to include this new aspect of the project.

CHAPTER 04

CONCLUSION

DESIGN
A PRACTICAL GUIDE TO RIBA PLAN OF WORK 2013
STAGES 2 AND 3

This book seeks to act as a practical guide to the RIBA Plan of Work 2013 Stage 2 Concept Design and Stage 3 Developed Design. It illustrates how to view the Plan of Work as a tool for organising, recording and monitoring tasks across the two stages and, together with the other two books in the series, the whole project process. The Plan of Work is intended to be flexible and able to be interpreted at an appropriate scale and complexity for the project at hand.

KEY LESSONS FROM THIS GUIDE

- At the outset of each of Stages 2 and 3, define what 'Concept Design' and 'Developed Design' mean respectively for this project.
- Is there a role for a client project champion? If so, who should that be?
- Take account of and review the procurement strategy, logging how decisions affect the types of information required within any stage.
- Review and update the Project Execution Plan (PEP), particularly after a procurement decision has been made.
- Introduce Change Control Procedure in the PEP in Stage 3.
- Review and update the Project Programme at both stages.
- Prepare a Design Programme for each design stage, with sufficient detail to drive consistent and collaborative decision-making.
- Review the validation requirements for a planning application, and ensure they are logged on the Design Programme.
- Include in the Design Programme field-study trips, design workshops and design team meetings, as agreed with the client.
- Use precedent studies to illustrate successful, relevant design elsewhere.
- Plan for an appropriate schedule of third party consultations, using a variety of types of engagement to produce the most constructive results.
- Review and update the Technology Strategy, including within it the BIM plan and referring to the BIM protocols expected for the project that are contained in the PEP.
- Reinforce the use of Project Strategies as a key component of managing the design process.
- Review the existing Project Strategies, and suggest others that should be established within either stage.
- Establish the level of detail required on Cost Information in each stage, and agree the basis on which costs are going to be prepared (eg historical data, square-metre rates, elemental costings).
- Review the sustainability checkpoints under each stage, and update the Sustainability Strategy accordingly.
- Plan the contents for the end of stage Information Exchanges, and share them with the design team at an early stage in order to streamline the production of information.

- Review at the outset the Stage 1 Communications Strategy, which clearly establishes how internal and external communications are to operate for the whole project. Build in a review cycle relating to Plan of Work stages.
- Encourage the design team to look ahead to activity in future stages, in order to give a purposeful context to the actions and decisions being taken in the current stage.

WHAT HAPPENS NEXT?

Once the client has reviewed the Stage 3 Information Exchange, and perhaps waited for planning permission to be granted, they will issue a permission to proceed to Plan of Work Stage 4 Technical Design. This stage produces technical information in drawn and written-specification form in order to allow construction to take place. As explained in this book, procurement choices can impact on when a cost for the project is fixed or when a contractor becomes involved in the processes leading up to the construction of the building, but on all projects it remains the case that the Stage 4 Technical Design needs to be complete on every element of the building design before Stage 5 Construction can take place.

DESIGN
A PRACTICAL GUIDE TO RIBA PLAN OF WORK 2013
STAGES 2 AND 3

Small residential extension for a growing family

Development of five new homes for a small residential developer

SCENARIO SUMMARIES

The three guide books in this series have used five scenarios to illustrate how different types and scales of project can all make successful use of the RIBA Plan of Work 2013 framework in order to plan and develop the design process, and to report progress to clients and other stakeholders along the way. A brief recap of what has occurred in these scenarios so far is included here up to Stage 4 Technical Design which will be covered in the next guide, Construction: A Practical Guide to the RIBA Plan of Work 2013 Stages 4, 5 and 6 by Phil Holden.

The client family appoints an architect who works with them to formulate an Initial Project Brief. A traditional form of contract is favoured. A Project Budget and Project Programme are drafted in order to set out the scope of the project.

A Concept Design is produced, along with a Sustainability Strategy.

A structural engineer is appointed to the design team, and a coordinated design is produced. The architect highlights the fact that not having a building services engineer on the design team is likely to cause problems later on, at tender stage.
Planning permission is achieved at the end of Stage 3.

The developer appoints an architect who prepares an Initial Project Brief and a Project Execution Plan in order to act as a control document for the project.

A structural engineer is appointed, and a site investigation commissioned.

The Concept Design is produced, along with a cost plan and developed Project Programme.

The client and architect undertake a number of visits to comparable schemes, to gauge the local market taste and price.

A planning application submission is made earlier than usual, at the end of Stage 2.

A traditional form of contract is the chosen procurement route. The procurement strategy highlighted this route as the one best able to cope with client variations and changes of mind, even if these impacted on costs.

CONCLUSION

Refurbishment of a teaching and support building for a university

The university commissioned an Initial Project Brief independently of the PQQ (pre-qualification questionnaire) process to appoint a design team.

A Concept Design is commenced, together with commissioning investigative survey work on the existing building. The Concept Design creates the possibility of additional teaching space to that envisaged in the Initial Project Brief. The university carry out an economic appraisal, and decide to develop the additional space.

The university's procurement strategy dictates that the contract will be let under a single-stage design and build form. The design team will most likely be novated at the end of Stage 3, but that decision will be taken nearer the time.

A Design Responsibility Matrix has been developed, indicating how the novation will affect design team activity.

During Stage 2, the university decided that the project would be a BIM pilot for them, and all design team members have committed to a collaborative BIM regime for the rest of the project.

A significant proportion of the tender documentation allows for contractor design portions of the construction.

At the end of Stage 3, no decision has been taken over whether the design team will be novated or not.

New central library for a small unitary authority

A Feasibility Study has been carried out, including a site options appraisal on council-owned sites.

Site investigations have been carried out on the chosen site, to help cost certainty for substructure in the cost plan.

The leader of the council has become the project champion.

The procurement strategy is to have a two-stage design and build contract, and the preferred contractor has worked collaboratively with the design team during Stage 3, making valuable commentary on the buildability of the library and, using prefabrication, has offered a significant reduction on the Stage 5 Construction period.

Change Control Procedures have been introduced, as the contract is to be let at the end of Stage 3.

New headquarters office for high-tech internet-based company

The client has high expectations of the building design, and, after a competitive-interview process, has selected a well-known architect and design team.

The Initial Project Brief has identified design quality and programme as the principal priorities, with value for money driving cost decisions.

The Stage 2 Concept Design has explored many options for this new headquarters building, including introducing the client to many similar precedent buildings in order to gauge their reaction to the design decisions demonstrated within them.

A project workshop early in Stage 3 reviewed the procurement strategy, and highlighted construction management as the most viable option to achieve the specific project requirements of speed and quality. A construction manager has been appointed early during Stage 3, and has estimated that about 15 packages will need to be tendered.

The construction manager has presented the client with a cost estimate within +3% of the Project Budget, which includes some limited market testing of the main packages and which has given the client the confidence to proceed.

A Developed Design was signed off by the client after the cost plan and Project Programme were agreed by the company board. The scheme was then submitted for planning permission at the end of Stage 3.

Due to the exceptional growth in employee numbers, the internet company have asked what the procedure would be for adding an extension of around 30% extra floor area as a contract variation.

The Handover Strategy has been developed during Stage 3 as the client wants to achieve an IT fit-out before Practical Completion, and has requested a Partial Possession of 25–35% of the floor area four months ahead of the completion predicted in the Project Programme.

WHY THE PLAN OF WORK FRAMEWORK IS
critical at the design stages

It's critical to use the Plan of Work framework at the design stages but it's also important to continue to follow the Plan of Work framework for the rest of the project. If the design team have successfully delivered the project to the end of Stage 3, it is useful to know what to expect from future stages. As mentioned above, the next guide in this series discusses Stage 4 Technical Design, Stage 5 Construction and Stage 6 Handover and Close Out. Stage 4 encompasses the completion of all of the pre-construction technical-design detailing and specifying. This information is generally used to competitively tender or to confirm a contractor's price for the project construction. The preferred contractor for the project plans their Construction Programme, site management and, of course, establishes their final contract price in order to enable a contract to be let. The level of detail required will depend on the Building Contract type, which, in turn, will have been controlled by the procurement strategy (see Task Bar 2) on the project-specific version of the RIBA Plan of Work 2013. After this, Stage 5 Construction can commence. The following stage, Stage 6, discusses the completion of the construction contract and the processes involved in the occupants starting to use and learning about how to operate the building.

It is crucially important for the design team, carefully managed by the lead designer, to comprehend the whole RIBA Plan of Work 2013 project framework, so that actions and decisions made during one stage of the process are taken with a clear understanding of how they will affect, adjust or assist in subsequent stages. Consciously planning for activity in future stages using the Design and Project Programmes allows judgements to be made about how long to leave options open for, and when to close options out in order to progress. Where priorities within the project criteria make it clear that a decision is essential within a stage, it is important to be clear about how the design team will arrive at an objective view – and how they will manage any consequences of that decision in future stages. The principal tools to assist the lead designer in these tasks are the Project Strategies. Well-compiled and comprehensive Project Strategies will plot a route through all of the Plan of Work stages from the very beginning of the project. The Information Exchanges during the design stages can communicate the developing Project Strategies with 'headline' outcomes and expectations for the current stage and each of the future stages. This process will mean that the design team can always check the current status of the project, but can also see ahead to the goals of each strategy in future stages.

Plan of Work glossary

A number of new themes and subject matters have been included in the RIBA Plan of Work 2013. The following presents a glossary of all of the capitalised terms that are used throughout the RIBA Plan of Work 2013. Defining certain terms has been necessary to clarify the intent of a term, to provide additional insight into the purpose of certain terms and to ensure consistency in the interpretation of the RIBA Plan of Work 2013.

'AS-CONSTRUCTED' INFORMATION

Information produced at the end of a project to represent what has been constructed. This will comprise a mixture of 'as-built' information from specialist subcontractors and the 'final construction issue' from design team members. Clients may also wish to undertake 'as-built' surveys using new surveying technologies to bring a further degree of accuracy to this information.

BUILDING CONTRACT

The contract between the client and the contractor for the construction of the project. In some instances, the **Building Contract** may contain design duties for specialist subcontractors and/or design team members. On some projects, more than one **Building Contract** may be required; for example, one for shell and core works and another for furniture, fitting and equipment aspects.

BUILDING INFORMATION MODELLING (BIM)

BIM is widely used as the acronym for 'Building Information Modelling', which is commonly defined (using the Construction Project Information Committee (CPIC) definition) as: 'digital representation of physical and functional characteristics of a facility creating a shared knowledge resource for information about it and forming a reliable basis for decisions during its life cycle, from earliest conception to demolition'.

BUSINESS CASE

The **Business Case** for a project is the rationale behind the initiation of a new building project. It may consist solely of a reasoned argument. It may contain supporting information, financial appraisals or other background information. It should also highlight initial considerations for the **Project Outcomes**. In summary, it is a combination of objective and subjective considerations. The **Business Case** might be prepared in relation to, for example, appraising a number of sites or in relation to assessing a refurbishment against a new build option.

CHANGE CONTROL PROCEDURES

Procedures for controlling changes to the design and construction following the sign-off of the Stage 2 Concept Design and the **Final Project Brief**.

COMMON STANDARDS

Publicly available standards frequently used to define project and design management processes in relation to the briefing, designing, constructing, maintaining, operating and use of a building.

COMMUNICATION STRATEGY

The strategy that sets out when the project team will meet, how they will communicate effectively and the protocols for issuing information between the various parties, both informally and at Information Exchanges.

CONSTRUCTION PROGRAMME

The period in the **Project Programme** and the **Building Contract** for the construction of the project, commencing on the site mobilisation date and ending at **Practical Completion**.

GLOSSARY

CONSTRUCTION STRATEGY
A strategy that considers specific aspects of the design that may affect the buildability or logistics of constructing a project, or may affect health and safety aspects. The **Construction Strategy** comprises items such as cranage, site access and accommodation locations, reviews of the supply chain and sources of materials, and specific buildability items, such as the choice of frame (steel or concrete) or the installation of larger items of plant. On a smaller project, the strategy may be restricted to the location of site cabins and storage, and the ability to transport materials up an existing staircase.

CONTRACTOR'S PROPOSALS
Proposals presented by a contractor to the client in response to a tender that includes the **Employer's Requirements**. The **Contractor's Proposals** may match the **Employer's Requirements**, although certain aspects may be varied based on value engineered solutions and additional information may be submitted to clarify what is included in the tender. The **Contractor's Proposals** form an integral component of the **Building Contract** documentation.

CONTRACTUAL TREE
A diagram that clarifies the contractual relationship between the client and the parties undertaking the roles required on a project.

COST INFORMATION
All of the project costs, including the cost estimate and life cycle costs where required.

DESIGN PROGRAMME
A programme setting out the strategic dates in relation to the design process. It is aligned with the **Project Programme** but is strategic in its nature, due to the iterative nature of the design process, particularly in the early stages.

DESIGN QUERIES
Queries relating to the design arising from the site, typically managed using a contractor's in-house request for information (RFI) or technical query (TQ) process.

DESIGN RESPONSIBILITY MATRIX
A matrix that sets out who is responsible for designing each aspect of the project and when. This document sets out the extent of any performance specified design. The **Design Responsibility Matrix** is created at a strategic level at Stage 1 and fine-tuned in response to the Concept Design at the end of Stage 2 in order to ensure that there are no design responsibility ambiguities at Stages 3, 4 and 5.

EMPLOYER'S REQUIREMENTS
Proposals prepared by design team members. The level of detail will depend on the stage at which the tender is issued to the contractor. The **Employer's Requirements** may comprise a mixture of prescriptive elements and descriptive elements to allow the contractor a degree of flexibility in determining the **Contractor's Proposals**.

FEASIBILITY STUDIES
Studies undertaken on a given site to test the feasibility of the **Initial Project Brief** on a specific site or in a specific context and to consider how site-wide issues will be addressed.

FEEDBACK
Feedback from the project team, including the end users, following completion of a building.

FINAL PROJECT BRIEF
The **Initial Project Brief** amended so that it is aligned with the Concept Design and any briefing decisions made during Stage 2. (Both the Concept Design and **Initial Project Brief** are Information Exchanges at the end of Stage 2.)

HANDOVER STRATEGY
The strategy for handing over a building, including the requirements for phased handovers, commissioning, training of staff or other factors crucial to the successful occupation of a building. On some projects, the Building Services Research and Information Association (BSRIA) Soft Landings process is used as the basis for formulating the strategy and undertaking a **Post-occupancy Evaluation** (www.bsria.co.uk/services/design/soft-landings/).

HEALTH AND SAFETY STRATEGY
The strategy covering all aspects of health and safety on the project, outlining legislative requirements as well as other project initiatives, including the **Maintenance and Operational Strategy**.

INFORMATION EXCHANGE
The formal issue of information for review and sign-off by the client at key stages of the project. The project team may also have additional formal **Information Exchanges** as well as

the many informal exchanges that occur during the iterative design process.

→ INITIAL PROJECT BRIEF

The brief prepared following discussions with the client to ascertain the **Project Objectives**, the client's **Business Case** and, in certain instances, in response to site **Feasibility Studies**.

→ MAINTENANCE AND OPERATIONAL STRATEGY

The strategy for the maintenance and operation of a building, including details of any specific plant required to replace components.

→ POST-OCCUPANCY EVALUATION

Evaluation undertaken post occupancy to determine whether the **Project Outcomes**, both subjective and objective, set out in the **Final Project Brief** have been achieved.

→ PRACTICAL COMPLETION

Practical Completion is a contractual term used in the **Building Contract** to signify the date on which a project is handed over to the client. The date triggers a number of contractual mechanisms.

→ PROJECT BUDGET

The client's budget for the project, which may include the construction cost as well as the cost of certain items required post completion and during the project's operational use.

→ PROJECT EXECUTION PLAN

The **Project Execution Plan** is produced in collaboration between the project lead and lead designer, with contributions from other designers and members of the project team. The **Project Execution Plan** sets out the processes and protocols to be used to develop the design. It is sometimes referred to as a project quality plan.

→ PROJECT INFORMATION

Information, including models, documents, specifications, schedules and spreadsheets, issued between parties during each stage and in formal Information Exchanges at the end of each stage.

→ PROJECT OBJECTIVES

The client's key objectives as set out in the **Initial Project Brief**. The document includes, where appropriate, the employer's **Business Case, Sustainability Aspirations** or other aspects that may influence the preparation of the brief and, in turn, the Concept Design stage. For example, **Feasibility Studies** may be required in order to test the **Initial Project Brief** against a given site, allowing certain high-level briefing issues to be considered before design work commences in earnest.

→ PROJECT OUTCOMES

The desired outcomes for the project (for example, in the case of a hospital this might be a reduction in recovery times). The outcomes may include operational aspects and a mixture of subjective and objective criteria.

→ PROJECT PERFORMANCE

The performance of the project, determined using **Feedback**, including about the performance of the project team and the performance of the building against the desired **Project Outcomes**.

→ PROJECT PROGRAMME

The overall period for the briefing, design, construction and post-completion activities of a project.

→ PROJECT ROLES TABLE

A table that sets out the roles required on a project as well as defining the stages during which those roles are required and the parties responsible for carrying out the roles.

→ PROJECT STRATEGIES

The strategies developed in parallel with the Concept Design to support the design and, in certain instances, to respond to the **Final Project Brief** as it is concluded. These strategies typically include:
- acoustic strategy
- fire engineering strategy
- **Maintenance and Operational Strategy**
- **Sustainability Strategy**
- building control strategy
- **Technology Strategy**.

These strategies are usually prepared in outline at Stage 2 and in detail at Stage 3, with the recommendations absorbed into the Stage 4 outputs and Information Exchanges.

The strategies are not typically used for construction purposes because they may contain recommendations or information that contradict the drawn information. The intention is that they should be transferred into the various models or drawn information.

GLOSSARY

QUALITY OBJECTIVES
The objectives that set out the quality aspects of a project. The objectives may comprise both subjective and objective aspects, although subjective aspects may be subject to a design quality indicator (DQI) benchmark review during the **Feedback** period.

RESEARCH AND DEVELOPMENT
Project-specific research and development responding to the **Initial Project Brief** or in response to the Concept Design as it is developed.

RISK ASSESSMENT
The **Risk Assessment** considers the various design and other risks on a project and how each risk will be managed and the party responsible for managing each risk.

SCHEDULE OF SERVICES
A list of specific services and tasks to be undertaken by a party involved in the project which is incorporated into their professional services contract.

SITE INFORMATION
Specific **Project Information** in the form of specialist surveys or reports relating to the project- or site-specific context.

STRATEGIC BRIEF
The brief prepared to enable the Strategic Definition of the project. Strategic considerations might include considering different sites, whether to extend, refurbish or build new and the key **Project Outcomes** as well as initial considerations for the **Project Programme** and assembling the project team.

SUSTAINABILITY ASPIRATIONS
The client's aspirations for sustainability, which may include additional objectives, measures or specific levels of performance in relation to international standards, as well as details of specific demands in relation to operational or facilities management issues.

The **Sustainability Strategy** will be prepared in response to the **Sustainability Aspirations** and will include specific additional items, such as an energy plan and ecology plan and the design life of the building, as appropriate.

SUSTAINABILITY STRATEGY
The strategy for delivering the **Sustainability Aspirations**.

TECHNOLOGY STRATEGY
The strategy established at the outset of a project that sets out technologies, including Building Information Modelling (BIM) and any supporting processes, and the specific software packages that each member of the project team will use. Any interoperability issues can then be addressed before the design phases commence.

This strategy also considers how information is to be communicated (by email, file transfer protocol (FTP) site or using a managed third party common data environment) as well as the file formats in which information will be provided. The **Project Execution Plan** records agreements made.

WORK IN PROGRESS
Work in Progress is ongoing design work that is issued between designers to facilitate the iterative coordination of each designer's output. Work issued as **Work in Progress** is signed off by the internal design processes of each designer and is checked and coordinated by the lead designer.

Index

Page numbers in italic indicate figures and in bold indicate glossary terms.

Nos
3D printing 120

A
accessibility strategy 110
acoustic strategy 61
acoustician 33
amenity and civic societies 108
approvals *see* planning permission
architect 16, 30, 33, 51, 70, 117
    see also lead designer
Arts Council England 43
'as-constructed' information **140**

B
BIM *see* Building Information Modelling (BIM)
biodiversity 98
budget, project 4, 119, **142**
building contract 94, 137, **140**
building façade design 61
Building for Life 12 (BFL12) 109
Building Information Modelling (BIM) 70, 85, 87, 118, **140**
Building Regulations 69
building services engineer 33, 70, 117
building services information 85–6
building users 110
business case **140**

C
CABE (Commission for Architecture and the Built Environment) 109
change control procedures 106–8, *107*, **140**
checklists, validation 97–8
civic societies 108
clash-detection software 85, 87

client 34–5, 44
co-located working 68
collaboration 68
common standards **140**
communication strategy 12, 101, **140**
compliance 102
    see also planning permission
concept design 4, *39*, 70, 86–7, *88*
    see also Stage 2 (Concept Design)
conservation architect 33
conservation area statements 98
construction programme 57, 137, **140**
construction strategy 11, 12, 61, *62*, 101, **141**
consultations
    public 108–9, 111, 112
    statutory 112
    third-party 108–12, *109*
contingency 52
contractor's proposals **141**
contractual tree **141**
coordinated design 85–6, 87
Core Objectives task bar 17, 18, 28, 80
cost consultant 33, 51, 53
cost estimates 52
cost information 51–2, 113, *114*, **141**
cost plan, elemental 52
costs
    designing to 51–4, 114
    presenting 54
    and procurement route 90
    project fees 19

D
Dance City, Newcastle upon Tyne 43
design and access statement 98
design and build procurement 56, 92, 93

INDEX

design coordination 85–6, 87
design development contingency 52
design leadership 14–16, 44
    see also lead designer
design process 13, 45, *45–50*, 50
design programme 19, 57–9, *59*, 95–6, *96*, 99, 119, **141**
design queries **141**
design responsibility matrix (DRM) 12, 19, 53, **141**
Design Review Panels 109
design team 14–16, *15*, 19
    Stage 2 4, 30, 32, 33, 36, 38
    Stage 3 7, 82, 87, 94
    see also architect; design workshops; lead designer
design team meetings 66, 95, *96*
design workshops 65–8, *67*, 110
designer, lead *see* lead designer
designing to cost 51–4, 114
designs
    concept 4, *39*, 70, 86–7, *88*
    developed 7, 84, *88*, 118
    see also Stage 2 (Concept Design); Stage 3 (Developed Design)
desktop site investigation 97
developed design 7, 84, *88*, 118
    see also Stage 3 (Developed Design)
digital technologies 120
drawings 45, *45–50*, 70, 117, 118
DRM *see* design responsibility matrix (DRM)

E
ecologist 33
ecology surveys 98
elemental cost plan 52
employer's requirements 93, **141**
end-of-stage cost models 52

F
Fabric First approach 69
façade engineer 33
feasibility studies **141**
feedback **141**
fees, project 19
field-study trips 41–3
final project brief 70, **141**
fire engineer 33
fire engineering strategy 61

G
geotechnical surveys 98
green travel plans 98

H
handover strategy 11, 12, *62*, **141**
health and safety strategy 11, 12, 61, *62*, 101, 119, **141**
heritage impact statements 98

I
information exchanges **141–2**
    at end Stage 1 20–1, 66
    at end Stage 2 6, 70–1
    at end Stage 3 8, 22, 94, 98, 116–19
Information Exchanges task bar 17, 28, 80
initial project brief 4, 21, 32, 35, 66, **142**
iterative design 50

K
key lessons 131–2

L
landscape architect 70, 117
lead designer 13, 14–16, 31, 33, 44
    Stage 2 tasks 38, 44, 51, 53, 57–8, 59, 61, 62
    Stage 3 tasks 92, 93, 94, 95
    see also architect

M
maintenance and operational strategy 11, 12, *62*, 101, 110, **142**
Malcolm Fraser Architects 43
management contracting 56, 92, 93
meetings, design team 66, 95, *96*
    see also design workshops
models, physical 45, *45*, *47*, 117, 118

N
noise surveys 98

O
operational strategy 111
outline cost plan 52

P
parametric design 120
PEP *see* project execution plan (PEP)
photo-real CGI 120

planning consultant 33
planning permission 60, 82–3, 94, 97–8, 112, 119
planning policy 98
post-occupancy evaluation (POE) **142**
practical completion **142**
pre-planning application process 83, 99
precedent projects 40–2, *40*
preliminaries 52
procurement route 51, 56, 90–4, *93*
procurement strategy 12, 55–6, 119
Procurement task bar 17, 18, 28, 80, 90
Programme task bar 17, 28, 80
project briefs
    final 70, **141**
    initial 4, 21, 32, 35, 66, **142**
project budget 4, 119, **142**
project champion 44
project execution plan (PEP) 21, 64, 70, 101, 119, **142**
project fees 19
project information **142**
project objectives 4, 12, 32, **142**
project outcomes 4, 12, 32, **142**
project performance **142**
project programme 4, 57–9, *57*, 119, **142**
project research 103–5
project roles table **142**
project strategies 4, 10–12, 137–8, **142**
    accessibility 110
    acoustic 61
    communication 12, 101, **140**
    construction 11, 12, 61, *62*, 101, **141**
    fire engineering 61
    handover 11, 12, *62*, **141**
    health and safety 11, 12, 61, *62*, 101, 119, **141**
    maintenance and operational 11, 12, *62*, 101, 110, **142**
    procurement 12, 55–6, 119
    research projects 103–5
    security 61, 111
    spatial quality 61
    Stage 1 12, 21
    Stage 2 12, 32, 61–2, *62*, *63*
    Stage 3 12, 87, 100–2, *100*, 119
    sustainability 11, 12, 61, *62*, **143**
    technology 12, 64, **143**
    third-party consultations 110–11
    tracker document 62, *63*

project team 33
    see also design team
provisional sums 52
public consultation 108–9, 111, 112

Q
quality objectives 32, **143**

R
research and development 36–7, 67, **143**
research projects 103–5
resource allocation 19
risk assessment 11, 37, *62*, 101, 102, **143**
risk registers 102
risk workshops 37

S
Scenario A (residential extension) 24, 74, 122–3, 134
Scenario B (small housing development) 24, 75, 123, 134
Scenario C (university building refurbishment) 25, 75–6, 86, 124, 135
Scenario D (new central library) 25, 76–7, 125, 135
Scenario E (new headquarters office) 25, 77, 86, 126, 136
schedule of services 19, **143**
Secured by Design Architectural Liaison Officer 111
security strategy 61, 111
sequential test assessments 98
services information 85–6
sign-off protocol 116–17
single-stage design and build 56, 92, 93
site information 4, **143**
site investigation, desktop 97
spatial design drawings 70
spatial quality strategy 61
specialist consultants 33
specifications 93
Stage 1 (Preparation and Brief) 4, 19, *93*
    information exchanges 20–1, 66
    project strategies 12, 21
Stage 2 (Concept Design) 27–77
    client expectations 34–5
    design process 13, 45, *45–50*, 50
    design programme 57–9, *59*
    design team 4, 30, 32, 33
    design workshops 65–8, *67*
    designing to cost 51–4

field-study trips 41–3
information exchanges 6, 70–1
iterative design 50
overview 4–6, *5*
planning permission 60
precedent projects 40–2, *40*
procurement 12, 55–6, *93*
project execution plan (PEP) 64, 70
project programme 57–9, *57*
project strategies 12, 32, 61–2, *62*, *63*
research and development 36–7, *67*
sustainability 69
task bars 16–18, 28, 69, 71
Stage 3 (Developed Design) 79–126
   change control procedures 106–8, *107*
   as completion of commission 94
   cost information 113, *114*
   describing the project 89
   design coordination 85–6, *87*
   design process 13
   design programme 95–6, *96*, 99, 119
   designing to cost 114
   developed design 7, 84, *88*, 118
   information exchanges 8, 22, 94, 98, 116–19
   overview 7–8, *9*
   planning permission 82–3, 94, 97–8, 112, 119
   procurement 90–4, *93*
   project strategies 12, 87, 100–2, *100*, 119
   research projects 103–5
   sign-off protocol 116–17
   sustainability 115
   task bars 16–18, 80, 90
   third-party consultations 108–12, *109*
Stage 4 (Technical Design) 133, 137
Stage 5 (Construction) 137
Stage 6 (Handover and Close Out) 137
statement of community involvement 98
statutory compliance 102
statutory consultations 112
strategic brief **143**
strategies *see* project strategies
structural engineer 33, 70, 117
structural information 85
Suggested Key Support Tasks task bar 17, 28, 80
sustainability 69, 115
sustainability aspirations 12, **143**
Sustainability Checkpoints task bar 17, 28, 69, 80
sustainability strategy 11, 12, 61, *62*, **143**

T

task bars 16–18, 28, 69, 71, 80, 90
technology strategy 12, 64, **143**
tendering 90, 92, 93
third-party consultations 108–12, *109*
time 95
town centre assessments 98
Town Planning task bar 17, 28, 80
tracker document 62, *63*
traditional procurement 56, 92, 93
transport assessments 98
transportation and highways engineer 33
tree surveys 98
two-stage design and build 56, 92, 93

U

UK Government Information Exchanges task bar 17, 28, 71, 80
users, building 110
utility providers 108

V

validation checklists 97–8
vibration surveys 98
virtual realities 120
visual impact assessments 98
visualisation techniques 120

W

work in progress **143**
workshops
   design 65–8, *67*, 110
   risk 37
   third-party consultations 110–11